# NEW APPROACHES TOWARDS A GRAND UNIFIED THEORY

Ray Munroe, Ph.D.

*Published by Lulu.com*

*Second Edition published May 2008*

*First Edition preprinted Nov. 2007, published Jan. 2008*

ISBN-13: 978-1-4357-0698-9

*This book is dedicated to my family, Lisa and Laura.*

# Contents

# Tables

## Figures

# 0. Abstract

This book will model the equilibrium quantum statistical mechanics of the Grand-Unified Mediating (GUM) boson that is implied by a quantum Grand-Unified Theory (GUT), and will yield the distinct low-energy force coupling strengths of that grand-unified field; thus explaining Quantum Statistical Grand Unified Theory (QSGUT) and thus showing that the graviton, photon, gluons, *W*'s and *Z* are all different forms of GUM Bosons. Additional forces and mass scales are predicted and presented as a test for experimentation. Finally, connections between this theory, Quantum Field Theory and Dirac's Large Numbers Hypothesis are explored; and underlying cosmological processes affecting Dark Matter, Dark Energy and Inflation are revealed. This book will also concentrate on the relationships between QSGUT, Group Theory (and Electroweak Theory), and String Theory (and Supersymmetry – SUSY). This Group Theory GUT model leads to five forces and a Hyperflavor extension of Electroweak Theory with new Hyperflavor fermions. And this SUSY GUT model leads to five generations of fermion matter particles with new Leptoquark fermions, and proposes new flavor-changing bosons that explain the CKM quark and PMNS neutrino mixing matrices.

05.90.+m, 12.10.-g, 12.10.Kt, 12.15.-y, 12.60.-i, 13.90.+i, 14.60.Pq, 14.80.-j

## 1. Introduction

For centuries, Physicists have striven towards the unity of similar ideas. In the 17th Century, Sir Isaac Newton unified Galileo Galilei's terrestrial gravity with Johannes Kepler's heavenly gravity to create a Universal Law of Gravitation [1]. In the 19th Century, James C. Maxwell unified electricity and magnetism with four equations representing Electromagnetism [2]. Great ideas by Albert Einstein, Bernhard Riemann and Theodor Kaluza could be interpreted as noble, yet incomplete, efforts to unify Gravitation and Electromagnetism in 4, 3 and 5 dimensions, respectively [3]. In the 1960's, Sheldon Glashow, Steven Weinberg and Abdus Salam unified Electromagnetism and the Weak Nuclear Force into Electroweak Theory [4]. And Quantum Field Theory with the Renormalization Group Equations implies a high-energy unification of the Strong and Electroweak (not Gravity) couplings [5], but the low-energy couplings seem arbitrary.

Traditionally, Particle Theorists call the search for One Theory a Grand Unified Theory (GUT), whereas String Theorists call it a Theory of Everything (TOE) – a presumptuous name. The quest for One Theory that would describe all of the four known forces: Gravitational, Electromagnetic, Strong and Weak Nuclear, has been hindered by the gravitational force. The other three forces are very different from gravity in that they are more important on nuclear and atomic scales where quantum mechanics plays a major role, whereas the much weaker gravity is more important on a cosmic (and less quantum influenced) scale because the other three forces are effectively charge shielded at these distances. But, regardless of the varying importance of quantum mechanics on the four known forces, all are expected to share the common field theory construct of requiring intermediate bosons to relay momentum and information between points.

## 2. A Thermodynamic Pattern

Science consists of finding patterns in data and trying to deduce an underlying process (whether we call it a hypothesis, theory or law) via understandable models. Table 1 summarizes an apparently thermodynamic pattern (because of the exponential dependence) in the low-energy coupling constants of the electromagnetic, weak and gravitational forces. This interesting pattern can be written as $\alpha_n = 1.8590 \times \exp\left(-0.34627 \times n^4\right)$ where $n$ = 1, 2, 3 and 4 are the low-energy coupling strengths of the strong, electromagnetic, weak and gravitational forces, respectively.

It seems appropriate that a Grand Unified Theory should involve Thermodynamics, the study of Energy. The Schrodinger equation of quantum

**Table 1 - A Thermodynamic Pattern**

| Force | S = Strength | ln (S) | Ratios of ln's vs. Quartic Integers |
|---|---|---|---|
| Strong Nuclear | $\sim 1$ | $\sim \pm 0$ | $\begin{cases} -4.9202/(\sim \pm 0) \\ \sim \mp\infty \neq 16. = 16/1 \end{cases}$ |
| Electromagnetic | $7.29735 \times 10^{-3}$ | $-4.92024$ | $\begin{cases} -26.517/(-4.9202) \\ = 5.39 \approx 5.06 = 81/16 \end{cases}$ |
| Weak Nuclear | $3.04562 \times 10^{-12}$ | $-26.5173$ | $\begin{cases} -88.025/(-26.517) \\ = 3.32 \approx 3.16 = 256/81 \end{cases}$ |
| Gravitational | $5.9060 \times 10^{-39}$ | $-88.025$ | |

mechanics contains a $\nabla^2\Psi$ term that leads to an integer-squared energy dependence for the solution of a particle in an infinite square well potential [6].

This quartic-integer (to-the-fourth power) dependence implies a $\nabla^4$ operator that is not a normal part of quantum mechanics, such as $E^4\Psi = \nabla^4\Psi$. This simple pattern misses the coupling strengths of the strong and weak forces by about a factor of three each, although this fit will be improved in upcoming Sections. Can an underlying theory reduce the number of apparently random "fundamental" constants?

## 3. A New Quantum Statistical Mechanics

The statistical mechanics of thermodynamics are based on the assumption that occupational probabilities are proportional to $\exp(-\beta E)$, although any time-invariant operator could be used as the basis for some type of statistical mechanics [7]. The energy-squared and quartic-energy (energy-to-the-fourth) are interesting cases, especially in relativity where energy values are given by $E^2 = m^2c^4 + p^2c^2$ and in particle physics where the Higgs field obtains a mass-squared with a negative vacuum expectation value (vev). Real masses are called tardons and must always travel slower than the speed of light in a vacuum, $c$; whereas imaginary masses are called tachyons and must always travel faster than $c$ (if the kinematics of special relativity also apply to tachyons). Our understanding of Spontaneous Symmetry Breaking (such as Higgs theory) implies that tachyons are virtual particles in a theory wherein the zero energy-squared state has been defined incorrectly, and that a correct definition of the zero energy-squared state will remove the tachyonic virtual particle and produce Spontaneous Symmetry Breaking.

Both energy-squared and quartic-energy are time invariant operators because, by Liouville's Theorem: $\partial E^m / \partial t = \left[H, E^m\right] = 0$, where $m$ = integer because the Hamiltonian is equal to (and commutes with powers of) the energy. Thus, either operator can become the basis of a legitimate type of statistical mechanics. The forms of the energy-squared and quartic-energy distribution functions for Maxwell-Boltzmann (MB) and Fermi-Dirac (FD) statistics are quite trivially the same as their respective thermodynamic distributions, but with energy variables replaced by energy-squared or quartic-energy variables, depending on the statistics.

What about the energy-squared and quartic-energy distribution functions for bosons? If the energy-squared weighting function for a particle is $\exp\left(-\beta^2 E^2\right)$, for two particles is $\exp\left(-\beta^2 E^2\right)\exp\left(-\beta^2 E^2\right)$ $=\exp\left(-2\beta^2 E^2\right)$, and so on, then the partition function for energy-squared Bose-Einstein (BE) statistics is a summable geometric progression:

$$Z=\sum_{m_n=0}^{\infty}\exp\left[-m_n\beta^2\left(E^2-\mu^2\right)\right]=\left\{1-\exp\left[-\beta^2\left(E^2-\mu^2\right)\right]\right\}^{-1}-1 \qquad (1)$$

The derivative of the partition function is also important:

$$-dZ/d\left[\beta^2\left(E_n^2-\mu^2\right)\right]=\sum_{m_n=1}^{\infty}m_n\exp\left[-m_n\beta^2\left(E_n^2-\mu^2\right)\right]$$
$$=\exp\left[-\beta^2\left(E_n^2-\mu^2\right)\right]\left\{1-\exp\left[-\beta^2\left(E_n^2-\mu^2\right)\right]\right\}^{-2} \qquad (2)$$

Dividing Equation (2) by Equation (1), we obtain the energy-squared Bose-Einstein distribution function:

$$f_{BE}\left[\beta^2\left(E_n^2-\mu^2\right)\right]=\langle m_n\rangle$$
$$=\sum_{m_n=1}^{\infty}m_n\exp\left[-m_n\beta^2\left(E_n^2-\mu^2\right)\right]\Big/\sum_{m_n=0}^{\infty}\exp\left[-m_n\beta^2\left(E_n^2-\mu^2\right)\right] \qquad (3)$$
$$=\left\{\exp\left[\beta^2\left(E_n^2-\mu^2\right)\right]-1\right\}^{-1}$$

And, recognizing that the quartic-energy or "$\mathcal{Q}$" for "$\mathcal{Q}$uartic" will have units of $(\text{Joules})^4$ but may not necessarily equal $\left[\operatorname{Re}\left(E^2\right)\right]^2$ (if, for example, the energy-squared has unobservable imaginary or matrix components), then the reasoning behind the quartic-energy Bose-Einstein distribution function follows Equations (1) through (3) and gives:

$$f_{BE}\left[\beta^4\left(\mathcal{Q}_n-\mu^4\right)\right]=\left\{\exp\left[\beta^4\left(\mathcal{Q}_n-\mu^4\right)\right]-1\right\}^{-1} \qquad (4)$$

# 4. Modeling and Preliminary Results

## 4.1 The Process - Quantum Statistical Grand Unified Theory

*Postulate 1: We assume that all force-carrying bosons are different states of the Grand-Unified Mediating (GUM) boson.*

The grand-unified force is sometimes mediated by massive bosons such as pions, $W$'s and $Z$'s; and is sometimes mediated by massless wave-particle bosons such as photons, gluons and gravitons. However, gluons are more fundamental than pions, thus a massless wave-particle boson model appears to be more fundamental than a massive boson model. The massless wave-particle boson model is also more universal because photons, gluons and gravitons will have the same basic characteristics whether they are mediating between tardons or tachyons; whereas pions, $W$'s and $Z$'s might require real mass bosons to mediate between tardons and imaginary mass bosons to mediate between tachyons.

Another complication is the conflict between the relativistic view of gravitation being a local continuous curvature of the spacetime manifold as opposed to the quantum view of gravitation being mediated locally by massless spin-two gravitons through a grainy spacetime. This book will approach gravitation using the latter view.

*Postulate 2: We assume that the coupling strengths of the fundamental forces are proportional to the respective density of states of the GUM boson, which can be modeled with the appropriate type of Quantum-Statistical Mechanics.*

Using quartic-energy Bose-Einstein statistics to model the density of states for GUM bosons will build a framework that will be suitable for both tardons and tachyons, and thus model known particles and the vev associated with the undiscovered Higgs boson. We will assume that the only variable that makes these different types of GUM bosons behave differently is their quantum number, *n*. The corresponding quartic-energy is denoted $\mathcal{Q}_n$.

In the massless wave-particle model, these GUM bosons are quanta with energy-momentum vectors of $hk_\alpha/2\pi$ and two allowable spin polarization states. If we further impose an ad-hoc condition in the form of periodic boundary conditions (such as Kaluza-Klein momentum or winding-mode excitations on a string [8] ) on our GUM bosons, then we will also have quantized momentum vectors. Thus, using a massless wave-particle model for the density of states for GUM bosons is similar to the problem of blackbody radiation [9], only done in quartic-energy intervals with quantized string momenta.

Imagine a blackbody cavity with D string membrane dimensions, each of length L, and at temperature T that is so dense that it is opaque to GUM boson radiation. Also assume that GUM boson division and recombination occurs at the walls of the cavity without any loss or gain of quartic-energy. This assumption allows us to approximate the quartic "chemical potential", $\mu^4$, as zero. There will be an infinite number of standing wave GUM boson modes with momenta $p_\alpha = n_\alpha hk_\alpha/2\pi$, and a uniform momentum space separation of $hk_\alpha/2\pi$ on a string membrane. The hyper-"surface area" that encloses a *D*-dimensional hypersphere of radius $r$ is given by $2\pi^{(D/2)}r^{(D-1)}/\Gamma(D/2)$, where $\Gamma(z)=(z-1)!$ with $\Gamma(1/2)=\sqrt{\pi}$ and $\Gamma(1)=1$ is the gamma function. This leads to a string momentum space density of states:

$$g(p)dp = \sum_{\alpha,\, spin}^{hypersurface} n_\alpha dn = 2 \times \left[ 2\pi^{(D/2)} (Lp/h)^{(D-1)} / \Gamma(D/2) \right] d(Lp/h)$$
$$= 4\left(\pi L^2/h^2\right)^{(D/2)} p^D \left(p^{-1}dp\right) / \Gamma(D/2) \tag{5}$$

Assuming periodic boundary conditions, such that $\mathcal{Q}_n = n^4 \mathcal{Q}_1 = n^4 (hc/2L)^4$, and using a change of variables, $\mathcal{Q} = p^4 c^4$, so that $\mathcal{Q}^{-1} d\mathcal{Q} = 4p^{-1}dp$, and transforming Equation (5) into a quartic-energy density of states gives:

$$\begin{aligned} g(\mathcal{Q})d\mathcal{Q} &= g(n, D, \mathcal{Q}_1)d\mathcal{Q} \\ &= 4\left(\pi L^2/h^2\right)^{(D/2)} \left(\mathcal{Q}/c^4\right)^{(D/4)} \left(\tfrac{1}{4}\mathcal{Q}^{-1}d\mathcal{Q}\right) / \Gamma(D/2) \\ &= \pi^{(D/2)} (L/h)^D (nh/2L)^D n^{-4} \mathcal{Q}_1^{-1} d\mathcal{Q} / \Gamma(D/2) \\ &= n^{(D-4)} (\pi/4)^{(D/2)} \mathcal{Q}_1^{-1} d\mathcal{Q} / \Gamma(D/2) \end{aligned} \tag{6}$$

Note that we are using Quantum Statistical Mechanics because our GUM bosons have integral quantized spins. This extra quantization condition, $\mathcal{Q}_n = n^4 \mathcal{Q}_1$, could be due to Kaluza-Klein momentum excitations on a string combined with $\mathcal{Q} = p^4 c^4$.

Now the number density of GUM bosons with quartic-energy values between $\mathcal{Q}_n - d\mathcal{Q}/2$ and $\mathcal{Q}_n + d\mathcal{Q}/2$ is given by:

$$N(n, D, \beta^4, \mathcal{Q}_1)d\mathcal{Q} = CM_n g(n, D, \mathcal{Q}_1) f_{BE}\left[\beta^4 \left(n^4 \mathcal{Q}_1 - \mu^4\right)\right] d\mathcal{Q} \tag{7}$$

where $C$ is a normalization constant, and $M_n$ is a degrees-of-freedom multiplicity factor to account for eight gluons versus one photon versus three massive Intermediate Vector Bosons $\left(IVB's = W^\pm, Z^0\right)$.

If we assume a large number of GUM bosons, and that the coupling strength of the $n^{\underline{th}}$ force in this non-interacting independent GUM boson model (denoted by $\alpha_n$) is proportional to the number density of GUM bosons in the $n^{\underline{th}}$ state, then:

$$\alpha_n = C_1 M_n n^{(D-4)} \Big/ \left\{ \exp\left[\beta^4 \left(n^4 \mathcal{Q}_1 - \mu^4\right)\right] - 1 \right\} \tag{8}$$

where $C_1(D)$ is a modified normalization constant containing dimensional factors from Equation (7). These generic GUM bosons condense to yield the properties of gluons, photons, IVB's or gravitons based on Thermodynamic occupation probabilities, and this is the fundamental reason for different valued couplings and charges.

### 4.2 Preliminary Results

Using the 2004 recommended values of fundamental physical constants [10], the fine structure constant is $\alpha_2 = e^2/2\varepsilon_0 hc = 7.297\ 352\ 568(24) \times 10^{-3}$. However, the coupling constants for the gravitational $(\alpha_4)$ and weak nuclear $(\alpha_3)$ forces are both mass dependent, and mass is not quantized. Thus, we must allow the masses to equal any reasonable mass (such as electron, pion or proton rest mass) and vary the number of string membrane dimensions, $D$, over integer values. Coincidentally, the best fit has dimension, $D = 3$ and gives the modeling results of Equation (8) in Table 2.

The modeled $\alpha_3$ is slightly less than the best value for weak leptonic (muon) decay expected at momentum transfers equal to the electron rest mass, $\alpha_3 = 8\pi^3 G_F m_e^2/(hc)^3 = 3.045\ 624(55) \times 10^{-12}$, but is very close to the expected value for weak nucleonic (neutron) decay – which is weaker because quark states mix.

**Table 2 – Results of Quantum Statistical Modeling of Force Strengths**

| $n$ | $n^{th}$ Force | $\alpha_n$ * | $n^{th}$ Charge | $n^{th}$ GUM Boson |
|---|---|---|---|---|
| 1 | Strong Nuclear | $69.021\ 75(55)$ | Color | 8 gluons |
| 2 | Electromagnetic | $7.297\ 352\ 568(24) \times 10^{-3}$ | Electric | 1 photon |
| 3 | Weak Nuclear | $2.948\ 07(12) \times 10^{-12}$ | Weak Isospin | 3 $W^{\pm}, Z^{0}$ |
| 4 | Gravitational | $5.906\ 03(88) \times 10^{-39}$ | Mass | 1 graviton |
| 5 | Fifth | $\sim 10^{-93}$ | ? | ? fifthons |
| 6 | Sixth | $\sim 10^{-193}$ | ? | ? sixthons |

*Digits in parenthesis indicate the propagated standard deviation in the last two digits of the given value. Here, $\mu^4 \approx 0$ is assumed, and $C_1 = 3.534\ 618(36)$ and $C_2 = \exp(-\beta^4 \mathcal{Q}_1) = 0.709\ 380\ 03(44)$ are obtained by fitting to the fine structure constant, $\alpha_2$, and the gravitational force strength between two protons, $\alpha_4 = G_N h m_p^2 c / 2\pi$.

If this modeled $\alpha_3$ represents nucleonic decays, then we gain information about the Cabibbo effect: $V_{ud} = 0.967\ 969(44)$ and $\theta_C \approx 14.66°$ (see correction in Section 5.2) versus the accepted $V_{ud} = 0.975\ 4(6)$.

Although this equilibrium Thermodynamic model ties all of the low-energy couplings together, and thereby implies a high-energy

Grand Unified Theory (GUT), the exact pathway to GUT should require non-equilibrium Thermodynamics, and leaves many questions unanswered (i.e. does the GUT contain partial GUT's such as $SU(2)_L \times U(1)_Y$ Electroweak unification, or is $SU(5)$ involved in GUT?). This simple model may also be modified by effects of a massive GUM boson ($\pi^0, \pi^\pm, W^\pm$ and $Z^0$ have masses of 0.1349743, 0.1395679, 80.39 and 91.1874 $\text{GeV}/c^2$, respectively), by non-linear effects (Renormalization), or by non-Abelian algebras such as $SU(3)_C, SU(2)_L \times U(1)_Y$, and $SU(2)_{I_3}$ [11] that allow GUM-GUM boson interactions.

This value for the strong coupling constant is nowhere near the High Energy Physicist's definition – the strength of a gluon field (~1), or the Nuclear Physicist's definition – the coupling strength between a nucleon and a meson field (~10 or more), and will be addressed later. Because there are an infinite number of quantized momenta states, there are also an infinite number of distinguishable forces in this model.

# 5. QSGUT Compared & Contrasted with Other Theories

The primary objective of this book, Quantum Statistical Grand Unified Theory (QSGUT), has been presented, but how does it relate to other theories? This QSGUT presumes to contain the Quantum Thermodynamics of Gravity, Electrodynamics, and the Nuclear forces. This chapter will attempt to forecast truths related to a Theory of Everything (TOE) by studying anticipated new physics in the overlapping regions between QSGUT and Quantum Field Theory.

## 5.1 Variability of Coupling Strengths

In 1937, Paul A. M. Dirac proposed the Large Numbers Hypothesis [12] (LNH), which was also pursued by Sir Arthur S. Eddington. One interpretation of the LNH proposes three related sets of dimensionless large numbers of order $10^{40}$ (the exponent ranges from 34 to 42, but 40 is a "nice round number") formed of ratios between cosmic-sized quantities: $\alpha_4^{-1}$, the size of the Universe, and the age of the Universe; and atomic-sized quantities: $\alpha_2^{-1}$, the Bohr radius, and a typical atomic transition time. The Large Numbers Hypothesis (LNH) concludes that large numbers are proportional, thus the Gravitational coupling is inversely proportional to the age of the Universe. But QSGUT says that a time-dependent $\alpha_4$ implies that $\exp\left(-\beta^4 \mathcal{Q}_1\right)$ and/ or $C_1$ are time-dependent, and none of the various couplings, sizes or times would be constants. Also, the ratios of the Weak Nuclear to the Fifth coupling strengths are of order "Large-Squared", i.e. $\alpha_3/\alpha_5 \sim 10^{80}$. These large and large-squared relationships between the couplings could be due to a causal GUT.

And in Particle Physics, Renormalizable Group Theory (RGT) leads to the Renormalization Group Equations (RGE's) that describe the variation with energy of the Strong and Electroweak couplings [13]. RGT is derived from Quantum Field Theory by renormalizing apparent singularities and summing over intermediate states via Feynman diagrams to obtain statistical averages. This QSGUT uses statistical averages from the first postulates, and thus abandons certain details available in RGT, but gains statistical insights that are not available with RGT. A generalization of the Correspondence Principle demands that overlapping observables between QSGUT and RGT must agree with each other if both theories are true representations of nature. Thus, the RGE's for the Electromagnetic and Weak forces should allow us to determine the variability of $\beta^4 \mathcal{Q}_1$ and $C_1$ and, therefore, $\alpha_4$ , with respect to a change in momentum transfer.

At low momentum transfer scales (renormalization $\mu$ much less than the mass of the $W$ boson), the electromagnetic and weak couplings are best given by Quantum Electrodynamics (QED) and the Fermi Weak Theory as opposed to $SU(2)_L \times U(1)_Y$ Electroweak Theory for larger values of $\mu$. Please note that the Standard Model RGT convention of $(\alpha_1, \alpha_2, \alpha_3)$ = (Electromagnetic $U(1)_Y$, Weak $SU(2)_L$, Strong $SU(3)_C$) has been converted into this QSGUT convention of $(\alpha_1, \alpha_2, \alpha_3)$ = (Strong, Electromagnetic, Weak), and this renormalization mass, $\mu$, (or momentum transfer $\mu c$) is not related to the quartic-energy chemical potential, $\mu^4$.

To one-loop, the RGE's give the variability in the Quantum Electrodynamic coupling as:

$$d(\ln \alpha_2)/d(\ln \mu) = -b_E \alpha_2 / 2\pi, \tag{9}$$

where $b_E = -\frac{4}{3}\sum_{f,c} q_{f,c}^2 = -\frac{32}{9}$ per generation, $q$ = electric charge, $f$ = fermion flavor and $c$ = color.

At momentum transfer scales far below the $W$ boson mass, the strength of the weak nuclear force is given by the Fermi theory, which is proportional to mass-squared:

$$\alpha_3 = 8\pi^3 G_F m^2 / (hc)^3 \text{, and } \mu^2 \propto m^2 \text{, so that } d(\ln\alpha_3)/d(\ln\mu) = 2. \tag{10}$$

Differentiating Equation (8), Quantum Statistical Grand Unified Theory gives:

$$\begin{aligned} &d(\ln\alpha_n)/d(\ln\mu) \\ &\quad = d(\ln C_1)/d(\ln\mu) - n^4\left[1 + (n\alpha_n / M_n C_1)\right] d(\beta^4 \mathcal{Q}_1)/d(\ln\mu) \end{aligned} \tag{11}$$

Combining Equations (9) through (11) gives the following results for momentum transfers far below the $W$ boson mass:

$$d(\beta^4 \mathcal{Q}_1)/d(\ln\mu) = -\left[2 + (b_E\alpha_2/2\pi)\right] / \left[65 - (32\alpha_2/C_1)\right] \approx -\frac{2}{65} < 0, \tag{12}$$

$$\begin{aligned} &d(\ln C_1)/d(\ln\mu) = \\ &\quad = -\left[32 + (81 b_E\alpha_2/2\pi) + (64\alpha_2/C_1)\right] / \left[65 - (32\alpha_2/C_1)\right] \\ &\quad \approx -\frac{32}{65} < 0, \end{aligned} \tag{13}$$

$$\begin{aligned} &d(\ln\alpha_{n\geq 2})/d(\ln\mu) = \\ &\quad = \left[2(n^4 - 16) + (n^4 - 81)(b_E\alpha_2/2\pi) + (2n^5\alpha_n / M_n C_1)\right. \\ &\quad \left. - (64\alpha_2/C_1) + (n^5 b_E\alpha_2\alpha_n / 2\pi M_n C_1)\right] / \left[65 - (32\alpha_2/C_1)\right] > 0 \\ &\quad \approx \begin{cases} -b_E\alpha_2/2\pi \text{ for } n = 2, \text{ and} \\ 2(n^4 - 16)/65 \text{ for } n \geq 3; \text{ thus} \end{cases} \end{aligned} \tag{14}$$

$$d(\ln\alpha_4)/d(\ln\alpha_2) \approx -192\pi/13 b_E\alpha_2 > 0. \tag{15}$$

Equation (15) says that changes in Dirac's coupling large number are proportional to a large number, which agrees with Dirac. If we also assume, as did Dirac, that the coupling large number $(\alpha_2/\alpha_4)$ is proportional to the time large number $(T/\tau_2)$, where $T$ is the age of the Universe $(\sim 14\text{ BY} \approx 4.4\times 10^{17}\text{ sec})$ and $\tau_2$ is a time scale for atomic transitions, then we can estimate the time variability of these "constants": $\alpha_4 = aT^{-1}$ where "$a$" is approximately constant. Although many studies have attempted to measure the time variability of these constants ($\alpha_2$ and $\alpha_4$), most do not consider the effects of correlated variable constants (Equation (15)). A star's luminosity [14] is approximately proportional to $(\text{mass})^4$, so a decreasing gravitational "constant" can cause distant stars (older signals) to act as if they are heavier and burn brighter than comparable nearby stars (younger signals). The frequency of a star's spectra is proportional to $\alpha_2^2$, so a decreasing fine structure "constant" can cause distant stars (older signals) to act as if they have a weakened Doppler redshift relative to nearby stars (younger signals).

When considered together, this gravitational effect will dominate this fine structure effect (from Equation (15) and the aforementioned power dependencies) causing old stars to be too bright, and modifying Hubble's Law (to be addressed later as Variable Coupling Theory, VCT). Spacetime may also be affected. Quantum Theory expects spacetime to become grainy at the Planck mass scale, $(hc/2\pi G_N)^{1/2}$. The ratio of the Planck mass to reasonable masses such as the proton or electron rest mass is approximately $10^{20}$, the square root of Dirac's large number (due to the $G_N^{-1/2}$ factor). Were Planck's mass and Dirac's large number both smaller in the early Universe? Dirac also predicted the existence of a magnetic monopole with a coupling constant of $\alpha_{MM} = 137/4$. This model does not include the magnetic monopole, but this should not be considered problematic because this particular particle has never been discovered and String/ M-Theory may be able to explain its relationship to electromagnetism via S-Duality arguments.

## 5.2 The Thermodynamic Properties of QSGUT

With the new physics that we obtained from the RGE's in the previous Section, we may now apply thermodynamic principles to get a better low-energy value for the strong coupling. In the Grand Canonical Ensemble, the thermodynamic equilibrium of a mixture is reached when the Gibbs free enthalpy (G) is minimized with respect to changes in composition [15], or $(\delta \mathrm{G})_{\mathrm{T,P}} = 0$. This represents a balance between minimized Helmholtz free energy (U + PV) and maximized entropy (−TS). Claude Shannon defines mathematical entropy as $-\sum \mathrm{p_i} \ln(\mathrm{p_i})$ where $\mathrm{p_i}$ is the fraction of a population in the $\mathrm{i}^{\underline{\text{th}}}$ state [16]. For normal thermodynamic systems, entropy can be defined as $-k_B \sum \mathrm{p_i} \ln(\mathrm{p_i})$. Here, we will assume that quartic-energy-based "thermo"-dynamic (hereafter called quartodynamic) entropy, $\mathcal{S}$, may be defined as $-k_B^4 \sum \mathrm{p_i} \ln(\mathrm{p_i})$. For this quartodynamic system, we will assume that the "Gibbs" Quartic-Enthalpy, $\mathcal{G}$, is of the form:

$$\mathcal{G} = \mathcal{U} + \mathcal{PV} - \mathcal{T}_{eff}^4 \mathcal{S} \tag{16}$$

where $\mathcal{U} = N \sum \mathrm{p}_n \left( \mathcal{Q}_n - \mu^4 \right)$, $\mathcal{PV} = -N \left( \mathcal{Q}_1 - \mu^4 \right)$,

$\mathcal{S} = -N k_B^4 \sum \mathrm{p}_n \ln(\mathrm{p}_n)$ and $\mathrm{p}_n = \alpha_n / \sum \alpha_n$

We will assume that this system is in equilibrium, $(\delta\mathcal{G})_{\mathcal{T},\mathcal{P}} = 0$, that constant $\mathcal{T}$ and $\mathcal{P}$ are equivalent to constant $\beta^4$ and $\mathcal{Q}_1$, and that we may simplify this problem to a two state system (strong nuclear and electromagnetic) with relative occupation probabilities in thermal equilibrium. If we only consider the strong nuclear and electromagnetic forces with relative occupation probabilities of $\mathrm{p}_1$ and $\mathrm{p}_2$, respectively, such that $\mathrm{p}_1 + \mathrm{p}_2 \approx 1$, then:

$$\begin{aligned} \mathcal{G}' &= N\sum_{n=1}^{\infty}\left[\left(n^4 - 1\right)\mathrm{p}_n\mathcal{Q}_1 + \beta^4\mathrm{p}_n \ln \mathrm{p}_n\right] \\ &\approx 15N\mathrm{p}_2\mathcal{Q}_1 + N\beta^4\left[\left(1-\mathrm{p}_2\right)\ln\left(1-\mathrm{p}_2\right) + \mathrm{p}_2 \ln \mathrm{p}_2\right] \end{aligned} \tag{17}$$

where we have used $\mathrm{p}_1 \approx 1 - \mathrm{p}_2$ and $\beta^{-4} \equiv k_B^4 \mathcal{T}_{eff}^4$.

Enforcing the equilibrium condition, $(\delta\mathcal{G})_{\beta^4,\mathcal{Q}_1} = 0$, we obtain:

$$\begin{aligned} (\delta\mathcal{G})_{\beta^4,\mathcal{Q}_1} = 0 \approx\ & N\mathcal{Q}_1\delta\mathrm{p}_2\left\{15 + \beta^{-4}\mathcal{Q}_1^{-1}\ln\left[\mathrm{p}_2/\left(1-\mathrm{p}_2\right)\right]\right\} \\ & + \mathcal{Q}_1\delta N\left\{15\mathrm{p}_2 + \beta^{-4}\mathcal{Q}_1^{-1}\left[\left(1-\mathrm{p}_2\right)\ln\left(1-\mathrm{p}_2\right) + \mathrm{p}_2\ln\mathrm{p}_2\right]\right\} \end{aligned} \tag{18}$$

This equation relates changes in the strong coupling to changes in the electromagnetic coupling (and weak coupling with the $\mathrm{p}_3$ term, see the general form of Equation (17)) and may have required the existence of quarks to carry both of these charges and help enforce this balance. Because both $C_1$ and $N$ are proportional to the number of GUM bosons, we will assume that $\delta(\ln N) \sim \delta(\ln C_1) \approx -32\left[\delta(\ln\mu)\right]/65$ from Equation (13). The variation of the occupation probability of photons, $\delta\mathrm{p}_2$, can be obtained from Equations (9), (16) and (22):

$$\delta\mathrm{p}_2 \approx \delta(\ln\mu)\left[b_S\alpha_2\left(1-\mathrm{p}_2\right)^2 - b_E\alpha_1\mathrm{p}_2^2\right]/2\pi \approx \delta(\ln\mu)\left(b_S\alpha_2\right)/2\pi \tag{19}$$

where we have simplified our result by assuming $1 >> p_2$ and $\alpha_2 \approx p_2 \alpha_1$. Also approximating $\ln(1-p_2) \approx -p_2$ and $-\ln(p_2) > 1 >> p_2$, and making the appropriate substitutions, our equilibrium condition now becomes:

$$\begin{aligned} 0 &= (\delta\mathcal{G})_{\beta^4,\mathcal{Q}_1} / N\mathcal{Q}_1[\delta(\ln\mu)] \\ &\approx b_S\alpha_2\left(15+\beta^{-4}\mathcal{Q}_1^{-1}\ln \mathrm{p}_2\right)/2\pi - 32\mathrm{p}_2\left[15+\beta^{-4}\mathcal{Q}_1^{-1}(-1+\ln \mathrm{p}_2)\right]/65 \end{aligned} \tag{20}$$

Substituting $b_S = 11 - \frac{2}{3}\sum n_f$, we obtain approximate low-energy equilibrium solutions for $\mathrm{p}_2$ and $\alpha_1$. Now we can apply the definition of the chemical potential (the "other $\mu$", not the renormalization energy) from the Grand Canonical Ensemble of Thermodynamics, $\mu = (\partial G/\partial N)_{T,P}$, and Equation (18) to this quartodynamic system:

$$\begin{aligned} \mu^4 &= (\delta\mathcal{G}/\delta N)_{\beta^4,\mathcal{Q}_1} = \\ &= \mathcal{Q}_1\left\{15\mathrm{p}_2 + \beta^{-4}\mathcal{Q}_1^{-1}\left[(1-\mathrm{p}_2)\ln(1-\mathrm{p}_2) + \mathrm{p}_2\ln \mathrm{p}_2\right]\right\} \end{aligned} \tag{21}$$

The simultaneous solution to our fit in Table 2 and Equations (20) and (21) gives $\mathrm{p}_2 = (4.693 - 0.084\, n_f) \times 10^{-3}$, $\mu^4 = -(1.6541 - 0.0063\, n_f) \times 10^{-2}\mathcal{Q}_1$ (an imaginary energy-squared!), $\exp(-\beta^4\mathcal{Q}_1) = 0.709\ 379\ 97$, $C_1 = 3.554\ 840 - (7.8 \times 10^{-5})\, n_f$, $\alpha_2 = e^2/2\varepsilon_0 hc$, $\alpha_3 = 2.948\ 12 \times 10^{-12}$ (or $V_{ud} = 0.967\ 986$), $\alpha_4 = G_N h m_p^2 c/2\pi$, $\alpha_1 = 68.0830 + 0.0036\, n_f$ (Equation (8)) and $\alpha_1 = 1.543 + 0.030\, n_f$ (from this Section's definitions and results – see the next Section for the resolution of this conflict). These parameterizations have been optimized for $2 \le n_f \le 4$ to accommodate renormalization energies of $\mu \le m_{Charm}c^2 \approx 1.2$ GeV.

### 5.3 The Strong Force and Renormalization

The value of the Strong Nuclear coupling varies over a wide range. This degenerate model does not include effects due to GUM-GUM boson interactions. In renormalizable QCD, gluon-gluon, gluon-quark-gluon, and etc. interactions contribute to the Green's functions that lead to the Renormalization Group Equations that govern the momentum-transfer dependence of the strong coupling. To one loop in the RGE's:

$$d(\ln \alpha_1)/d(\ln \mu) = -b_S \alpha_1 / 2\pi, \tag{22}$$

where $b_S = 11 - \frac{2}{3}\sum_{f=u,d}^{t,b} n_f$, and $n_f =$ number of quark flavors, so that

$b_S\left(\mu = m_{Strange}c^2\right) = 9.$

A program by Ian Hinchliffe combines 2001 Review of Particle Properties Data with the RGE running couplings to obtain the best fit values for the Strong Nuclear coupling for a given renormalization energy [17]. This program gives $\alpha_1 = 1.63$ for $\mu = 585^{+58}_{-54}$ MeV or $\alpha_1 = 68.1$ for $\mu = 416^{+41}_{-39}$ MeV. Note that both renormalization energies, $\mu = 416$ and $585$ MeV, are on the order of the $K^+$ meson mass of $493.7$ MeV / $c^2$ (the lightest strange flavored meson), and thus have $n_f \approx 3$, and this was assumed to obtain a value of $\alpha_1 = 1.54 + 0.03\, n_f = 1.63$. Thus, a variance in the renormalization mass scale by a mere 30% can explain the differences between $\alpha_1 = 1.63$ and $\alpha_1 = 68.1$.

What physical phenomena are occurring at the scale of GUM bosons? These gluons are Bose-Einstein Condensates of the generic GUM bosons. The fact that so many GUM bosons are in the gluon state proves that this Bose-Einstein distribution is at a low effective temperature. Whereas a Bose-Einstein Condensate can have interesting macroscopic properties such as Superfluidity, gluons partially counteract these properties by obeying Color Confinement. Thus, the Bose-Einstein Condensate tries to attract more GUM bosons into the gluon state, but these excess gluons are then bound up into hadrons, mesons or glueballs. The net effect is that more gluons do not lead to more "free" (non-color-bound) gluons. The effects of color confinement are mathematically expressed in the Renormalization Group Equations (specifically, the two loop version of Equation (22)).

### 5.4 A Hierarchy of Masses

The Standard Model of Particle Physics includes a Higgs field [18] that obtains a negative vacuum expectation value (vev – a negative energy-squared) of order 246 GeV and a Higgs boson of similar mass that gives masses to other particles via its interactions with them. This Higgs boson should also couple to the Planck or GUT mass scale and these radiative corrections should drive this mass to a much larger value, which is the Gauge (or Mass) Hierarchy Problem. The Minimal Supersymetric Standard Model (MSSM) solves this Hierarchy Problem by introducing supersymetric particles that complete the circle of radiative corrections and allow a stable, lightweight Higgs boson [19]. The MSSM complicates the Higgs boson sector that now includes light $H_L$, heavy $H_H$, pseudoscaler $H_P$, and charged $H^{\pm}$ Higgs bosons.

The field of Solid State Physics has a Semiclassical Model of electron dynamics whereby a Bloch electron within a crystalline lattice can obtain an effective mass tensor that is very different from an electron rest mass due to interactions with the lattice. Probing photons or electrons with momentum vector, $hk/2\pi$, can measure these electron band characteristics. Positive masses are electrons and negative masses are holes (the lack of an electron with opposite effective charge). If the curvatures of the electron energy bands are relatively small, then we can make the semiclassical approximation [20]:

$$m_n^{*\,-1} = \left(\frac{2\pi}{h}\right)^2 \partial^2 E_n(k)/(\partial k)^2 \tag{23}$$

where $E_n(k)$ describes the electron/ hole energy band and $m_n^*$ is the effective mass. This model is based on $E = \frac{p^2}{2m}$, so that $\frac{\partial^2 E}{\partial p^2} = m^{-1}$, and substitute $\mathbf{p}\Psi = -i\,\nabla_x \Psi = \frac{h\mathbf{k}}{2\pi}\Psi$.

Earlier, we used a massless blackbody radiation model for GUM bosons and ignored massive GUM bosons, however a similar effective mass concept may apply to this theory. We will square all terms to make this equation look more relativistic, and replace the momentum vector, $hk/2\pi$, with the momentum transfer, $\mu c$, from earlier:

$$m_n^{*\,-2} = \partial^2 E_n^2(\mu)/\partial\left(\mu^2 c^2\right)^2 \tag{24}$$

where $E_n^2(\mu) = n^2\left(\alpha_n \mathcal{Q}_1\right)^{1/2}$ describes the variation of our "energy bands".

There are two special cases because of different forms of QSGUT-derived RGE's for these forces:

$$d(\ln \alpha_{n\le 2})/d(\ln \mu) = -\,b_n \alpha_n/2\pi, \text{ and} \tag{25}$$

$$d(\ln \alpha_{n\ge 3})/d(\ln \mu) = 2\left(n^4 - 16\right)/65 \tag{26}$$

from Equations (9), (14) and (22). We can use the above equations and a modified Equation (12): $d\mathcal{Q}_1/d(\ln\mu) \approx -2C_3/65\beta^4$ and $d\beta^4/d(\ln\mu) \approx 2(C_3-1)/65\mathcal{Q}_1$, such that $d(\beta^4\mathcal{Q}_1)/d(\ln\mu)$ $=(\beta^4 d\mathcal{Q}_1 + \mathcal{Q}_1 d\beta^4)/d(\ln\mu) = [-2C_3 + 2(C_3-1)]/65 = -2/65$.

Taking the derivatives and keeping leading terms gives the following results:

$$m^{*\,-2}_{n\le 2} = n^2(\alpha_n\mathcal{Q}_1)^{1/2}/(\mu c)^4\{3(b_n\alpha_n/8\pi)^2 + (b_n\alpha_n/8\pi)\times[1 + C_3/65\,\beta^4\mathcal{Q}_1] \tag{27}$$

$$+ C_3/(130\,\beta^4\mathcal{Q}_1) - C_3^2/(130\,\beta^4\mathcal{Q}_1)^2\} \tag{28}$$

$$m^{*\,-2}_{n\ge 3} = n^2(\alpha_n\mathcal{Q}_1)^{1/2}/(\mu c)^4\{[(n^4-16)/130]^2 - [(n^4-16)/130]\times[1 + C_3/65\,\beta^4\mathcal{Q}_1] \tag{29}$$

The term in Equation (28) must be added to both Equations (27) and (29). Note that $b_2\alpha_2/8\pi$ and $-(3^4-16)/130$ are both relatively large negative vev terms, and Higgs theory requires at least one. Table 3 is a fit to the pion mass and Higgs vev scales and yields $C_3 = 4.6012\times10^{-2}$ (and thus $\mathcal{Q}_1 d\beta^4/d(\ln\mu)$ dominates $\beta^4 d\mathcal{Q}_1/d(\ln\mu)$) and $C_4 = (\mu c)^4/\mathcal{Q}_1^{1/2}$ $=(201.48\ \text{MeV}/c^2)^2$. This raw fit missed the Planck scale by a factor of 140, but this is a semiclassical, not a quantum-gravity, model. In the next column, an ad-hoc correction term of $\alpha_n^{-0.0458}$ (which brings the power of $\alpha_n$ down from 0.5 to 0.4542 in Equations (27) through (29) – also, the above quoted values for $C_3$ and $C_4 = (\mu c)^4/\mathcal{Q}_1^{1/2}$ are after this correction) has been applied to bring the corrected value of $m_5^*$ down to the expected value.

**Table 3 - A Heirarchy of Masses**

| $m_n^*$ | Raw Mass $\left(\mathrm{GeV}/c^2\right)$ | Corrected Mass $\left(\mathrm{GeV}/c^2\right)$ | Mass Scale |
|---|---|---|---|
| $m_1^*$ | 0.134974 | 0.134974 | Pion mass scale |
| $m_2^*$ | 246 ***i*** | 246 ***i*** | Negative Higgs vev? |
| $m_3^*$ | 104 ***i*** | 55.8 ***i*** | New negative vev scale? |
| $m_4^*$ | $1.5\times10^{8}$ | $1.9\times10^{7}$ | New generations & Gravity GUT? |
| $m_5^*$ | $1.6\times10^{21}$ | $1.2\times10^{19}$ | Planck mass scale |

This gives a correction term of similar order as $C_3$ and $\alpha_2$:

$$m_{n\,\mathrm{corr}}^{*\;-2} = m_n^{*\;-2}\alpha_n^{\varepsilon} \text{ with } \varepsilon = 4.58\times10^{-2}. \tag{30}$$

At this point, it is appropriate to revert to the Large Numbers Hypothesis, and argue that the Planck mass is the square root of a large number, and therefore we should expect mass-squared to be inversely proportional to the coupling. However, this is not what Equations (27) through (29) say. In this case, mass-squared is inversely proportional to the square root of the coupling, and this fit only works because $\alpha_5^{-1}$ is of order Large-Squared. The "derivation" of Equation (24) assumes a classical limit $\left(E=\dfrac{p^2}{2m}\right)$ and relies on squaring all terms, and there is no good reason for this step other than that it yields negative vev's.

References [5] imply that the Standard Model (SM) RGE's with no new physics is incompatible with a GUT, but that if we assume new physics at the ~ TeV energy scale (usually SUSY), then we can obtain a GUT at the $\sim10^{13}$ TeV energy scale. The justification for a GUT at $\sim10^{13}$ TeV rather than the Planck scale of $\sim 10^{16}$ TeV is questionable. This Section's results imply that new physics (possibly new force bosons, new generations of matter fermions, and/ or SUSY?) at the $\sim 10^{4}$ TeV energy scale lead to a GUT at the Planck scale. The Mass Hierarchy problem now becomes a fine-tuning issue of 1 part in $\left(10^{4}\text{ TeV}/0.1\text{ TeV}\right)^{2} \sim 10^{10}$ (the ratio of the electromagnetic to weak coupling – this is much smaller than the non-SUSY SM fine tuning of 1 part in $\sim 10^{34}$ ~ Dirac's Large Number) if SUSY is on the scale of $m^{*}_{4\,\text{corr}}$. However, it may be that there are only these five stable masses, and all real masses are "small" (perhaps $\leq \left|m^{*}_{2\,\text{corr}}\right|$) radiative corrections about these stable points.

In Standard Electroweak theory, the Electromagnetic and Weak Nuclear forces mix with the Weinberg (or weak) mixing angle, $\theta_W$. Is it possible for some of these energy bands (two bands represent negative mass-squared bosonic "holes" and the other bands represent positive mass-squared bosonic "electrons" – using Solid State Physics analogies) to mix and produce new quasiparticle excitations? A GUT Lie algebra, such as $SU(5)$ with three degenerate Strong mass states ($3\ \text{m}_1$'s yielding $\pi^0$, $\pi^{\pm}$), one Electromagnetic mass state ($\text{m}_2 \neq \text{m}_\gamma$), and one Weak mass state ($\text{m}_3$), might allow these inverse-mass-squared components to mix and yield new states. These quasiparticles might have effective masses given by:

$$\left(m_{1\,eff}^{-2}, m_{2\,eff}^{-2}, \dots, m_{N\,eff}^{-2}\right) = \begin{pmatrix} a_{11} & \cdots & a_{1N} \\ \vdots & SU(N) & \vdots \\ a_{N1} & \cdots & a_{NN} \end{pmatrix} \begin{pmatrix} m^{*\,-2}_{1\,\text{corr}} \\ \vdots \\ m^{*\,-2}_{N\,\text{corr}} \end{pmatrix} \tag{31}$$

What happens when one of the $m_{n\,eff}^{-2}$ or $m_n^{*\,-2}$ is zero? Note that Equations (27) through (29) for $m_2^{*\,-2}$ and $m_3^{*\,-2}$ both contain relatively large negative terms, and either term or one of the $m_{n\,eff}^{-2}$ could have equaled zero. Could there have been a time in the history of the Universe, perhaps momentarily after the Big Bang, when the $W^{\pm}$ or $Z^{0}$ masses approached $\pm\infty$? In reality, the total Energy – Mass of the system could not have reached infinity, but quantum effects could have allowed tunneling past this apparent infinity. Immediately after tunneling through this potentially stable infinity, the strong force would have strengthened while all other forces greatly weakened, causing Alan Guth's Inflationary period as gravity greatly loosened its grip on the expanding mass of the Universe [21]. There may have been more than one apparently stable infinities (due to $m_2^*$, $m_3^*$ and / or $m_{n\,eff} \to \pm\infty$) and Inflationary periods in the early Universe.

As astrophysicists study the rotation rates of galaxies, we find that most of the mass of a galaxy is non-luminous "Dark Matter", thus implying a significant amount of matter that is different from us and our known solar system, whether this mass is from Black Holes, MACHO's (Massive Compact Halo Objects), WIMP's (Weakly-Interacting Massive Particles), or etc. The Lightest Supersymmetric Particle (LSP) is expected to be a stable WIMP. This new supermassive mass scale, $m_4^*$, is large enough that any new supermassive particles might be WIMP Dark Matter.

Up to this point, the RGE's and Dirac's Large Number Hypothesis combined with this theory have implied that these fundamental "constants" changed in the past, but are they still changing today? We should expect the Energy-Mass dynamics of this system to be stable against variations in $\mu$ if the following conditions apply:

$$\partial\left(E^2 + m^2c^4\right)/\partial\mu = 0 \text{ and } \partial^2\left(E^2 + m^2c^4\right)/\partial\mu^2 > 0 \text{ at } \mu = \mu_0 \text{ or}$$
$$\partial\left(E^4 + m^4c^8\right)/\partial\mu = 0 \text{ and } \partial^2\left(E^4 + m^4c^8\right)/\partial\mu^2 > 0 \text{ at } \mu = \mu_0 \qquad (32)$$

where $E^2 = \sum_n \alpha_n n^2 \mathcal{Q}_1^{1/2}$, $m^2 = \sum_n \alpha_n m_{n\,\mathrm{corr}}^{*\,2}$, $E^4 = \sum_n \alpha_n n^4 \mathcal{Q}_1$ and $m^4 = \sum_n \alpha_n m_{n\,\mathrm{corr}}^{*\,4}$.

Should we use energy-squared and mass-squared, or quartic-energy and quartic-mass for this minimization? A simple order-of-magnitude analysis demands that stronger forces will contribute more, and weaker forces will contribute less to a possible stable equilibrium at $\mu = \mu_0$. Ultimately, this implies that the pion mass largely defines any equilibrium of this system, and the pion mass helps to prevent the dismal disintegration of the Universe's force couplings into a very strong Strong Nuclear force and very weak other forces. If the system overshot this minimum, it should oscillate around $\mu = \mu_0$.

For energy-squared and mass-squared weighting, there are no stable minima. For quartic-energy and quartic-mass weighting, there is a stable minimum for $\mathcal{Q}_1^{1/4} = 199.0\ \mathrm{GeV}$ which simultaneously requires that the thermal energy $\beta = (260.0\ \mathrm{GeV})^{-1}$, the renormalization mass $\mu = 6.332\ \mathrm{GeV}/c^2$ (more than ten times the results of Section 5.3 – Have we failed to consider important correction terms such as Equation (22) being too poor an approximation of the Strong Nuclear RGE which impacts the pion mass, or has the Universe not yet reached equilibrium?), and the magnitude of the chemical potential $|\mu| = 71.24\ \mathrm{GeV}$. At this minimum, the weighted sum of mass-squared is negative (negative Higgs vev's?), and the weighted sum of quartic-mass is only 1% of the weighted sum of quartic-energy (Dark Energy?). All of these quantities might have values between the pion mass scale $(135.0\ \&\ 139.6\ \mathrm{MeV}/c^2)$ and the Higgs vev scale $(\text{of order } 246\ \mathrm{GeV})$, as we should expect from fine-tuning arguments.

## 5.5 The Fifth Force - WIMP-Gravity

Our distribution allows the possibility of an infinite number of quantized GUM boson states. However, we need to know the representation groups for gravity and all higher-*n* forces to determine $M_n$. Quantum gravity should be mediated by a rank-two tensor boson with zero rest mass.

Clifford algebra may have an application to this question. Clifford algebra of level four (see Table 4) contains Hamilton's quaternions and Dirac's $\gamma$ matrices. Maxwell's Equations of Electromagnetism can be written as one quaternion equation. And Dirac's $\gamma$ matrices are useful with Lorentz transformations and the Dirac Equation. Clifford algebra of level four also contains space-time transformations, and may be decomposed as one each scalar and pseudo-scalar, four each polar and axial vectors, and six anti-symmetric rank-two tensors. These six rank-two tensors may be represented as *SO*(4), which is equivalent locally to $SU(2)\times SU(2)$. However, *SU*(2) is a non-Abelian Lie algebra, and we should expect the group containing gravity to obey an Abelian Lie algebra (if the graviton has a non-zero rest mass and couples to mass, then it cannot mediate an infinite-range force such as gravity), therefore this *SO*(4) of tensor bosons cannot represent gravity as we know it.

Although, it is interesting to note that a complex Clifford algebra of level four is equivalent to a real Clifford algebra of level five, and the Clifford algebras of levels five and six contain ten (an *SO*(5) rank-two tensor which is equivalent locally to an *Sp*(4)) and fifteen (an *SO*(6) rank-three tensor which is equivalent locally to an *SU*(4)) components, respectively. The Einstein field equations of General Relativity [22] are 10 independent rank-two tensor equations resembling the *SO*(5)'s in Clifford algebra of level five.

**Table 4 - A Pictorial Representation of Select Clifford Algebras**

| Level | Order | Application |
|---|---|---|
| 0 | 1 | Real Numbers |
| 1 | 1 1 | Complex Numbers |
| 2 | 1 2 1 | Pauli $\sigma$ Matrix Algebra |
| 3 | 1 3 3 1 | Vector Algebra |
| 4 | 1 4 $6^{A2}$ 4 1 | Dirac $\gamma$ Matrix Algebra & Quaternions |
| 5 | 1 $\underline{5}$ $\underline{10}^{S2}$ $\underline{10}^{A2}$ $\underline{5}$ 1 | Einstein Field Equations? |
| 6 | 1 6 $\underline{15}^{S3}$ $\underline{20}^{A4}$ $\underline{15}^{A3}$ 6 1 | |
| | ↑ Sc ↑ PV ↑ AV ↑ PS | |

where Sc = Scalar, PV = Polar Vector, S2 = Symmetric Rank-2 Tensor, A2 = Antisymmetric Rank-2 Tensor, A3 = Antisymmetric Rank-3 tensor, AV = Axial Vector, PS = Pseudoscalar, etc.

Now suppose that the *SO*(4) of anti-symmetric rank-two tensors represented gravity during the assumed early period of Grand Unification (just after the Big Bang) and the *SO*(5) of anti-symmetric rank-two tensors represented a related force that the author calls Weakly Interacting Massive Particle Gravity or WIMP-Gravity. These force quanta could have mixed quantum numbers and broken spontaneously into an Abelian *U*(1) algebra of gravity and a non-Abelian *SO*(6) or *SU*(4) algebra of WIMP-Gravity (similar to Electroweak Theory, but with different algebra groups – the massless graviton is thus created!). Such an *SO*(6) group could have further broken into an $SO(3)\times SO(3)$ algebra group of real and imaginary three dimensional spaces (which is equivalent locally to $SU(2)\times SU(2)$) – this is a particularly likely scenario if WIMP-Gravitons are massive, couple to mass, and interact with tardons and tachyons alike.

Finally, if we assume an *SO*(3) Lie algebra group for real (tardonic) WIMP-Gravity, then we can update Table 2 by stating that the Fifth force is WIMP-Gravity, is mediated by three very massive WIMP-Gravitons designated by "$F$", and has a relative coupling strength of $1.34\times 10^{-93}$ for small momentum transfers (on the proton mass scale). If we assume that the interaction strength of gravity is proportional to $(\text{mass})^2$ and WIMP-Gravity is proportional to $(\text{mass})^{2j}$, $j$ = integer, then we can obtain a unification of Gravity and WIMP-Gravity below the Planck mass scale of $m_{Planck}\sim 1.2\times 10^{19}$ GeV / $c^2$ if $j \geq 3$. Note that Equation (14), which defines the variation of these couplings with momentum transfer, is only valid below the Electroweak scale.

Because WIMP-Gravity may involve imaginary masses (with $i^2=-1$, and $i^4=1$), a reasonable assumption would be that the interaction strength of WIMP-Gravity is proportional to $(\text{mass})^8$ (where 4 = $j$ ≥ 3), thus allowing the unification of Gravity and WIMP-Gravity at a momentum transfer scale of $(\alpha_4/\alpha_5)^{1/(8-2)}m_p c\sim 1.2\times 10^9$ GeV / $c$, where $m_p$ is the mass of the proton; and thus implying a WIMP-Graviton mass of $\sim 1.2\times 10^9$ GeV / $c^2$ (60 times $m^*_{4\,\text{corr}}$, see Table 3). We will assume that Equation (14) specifies the renormalization mass dependence of Gravity and WIMP-Gravity at low energy scales (below an $E_x$ to be determined), that this paragraph's power assumptions specify these mass dependences at higher energy scales (above $E_x$), and that Gravity and WIMP-Gravity unify at the mass scale of $m^*_{4\ \text{corr}} = 1.9 \times 10^7\ \text{GeV}/c^2$. Then we can determine that $E_x = 97.9\ \text{GeV}$, and the unified coupling strength of Gravity and WIMP-Gravity would be $\alpha_{Gravity} = 1.8 \times 10^{-13}$. We can verify this unification as follows:

$$5.9 \times 10^{-39} \times (97.9\ \text{GeV}/0.938\ \text{GeV})^{7.38} \times \left(1.9 \times 10^7\ \text{GeV}/97.9\ \text{GeV}\right)^2$$

$$= 1.3 \times 10^{-93} \times (97.9\ \text{GeV}/0.938\ \text{GeV})^{18.7} \times \left(1.9 \times 10^7\ \text{GeV}/97.9\ \text{GeV}\right)^8$$

$= 1.8 \times 10^{-13}$ (using $2\left(n^4 - 16\right)/65 = 7.38$ for $n = 4$ or $18.7$ for $n = 5$). At the Gravity/ WIMP-Gravity unification scale, the interaction/ decay time for such a WIMP-Graviton would be

$$t_{Gravity} \sim t_3 \alpha_3 \alpha^{-1}_{Gravity} \sim \left(10^{-10}\text{s}\right) \times \left(3.0 \times 10^{-12}\right) \times \left(1.8 \times 10^{-13}\right)^{-1} \sim 1.7 \times 10^{-9}\text{s},$$

where $t_3$ is a typical interaction/decay time associated with the Weak Nuclear Force. Thus, these somewhat unstable, primordial WIMP-Gravitons should have survived the first $\sim 10^{-35}$ s with the early periods of GUT's and Inflation. Based on current coupling values, $\alpha^{-1}_{Gravity} = \alpha_4^{-1} = 1.6 \times 10^{40}$, and $t_{Gravity} \sim 5 \times 10^{18}\ \text{s} \sim 150$ Billion Years, and any WIMP-Gravitons that have survived to the present epoch should be "frozen out" in the sense that they should not easily interact with anything. It is interesting to note that the highest cosmic ray energies [23] yet to be detected have energies of about $3 \times 10^{11}\ \text{GeV}$, and the same mechanisms that are capable of producing these powerful cosmic rays may also be able to produce WIMP-Gravitons, or these powerful cosmic rays may be caused by WIMP-Gravitons interacting with strong gravitational fields.

**Figure 1 – From GUT To GUM Boson Quantum States**

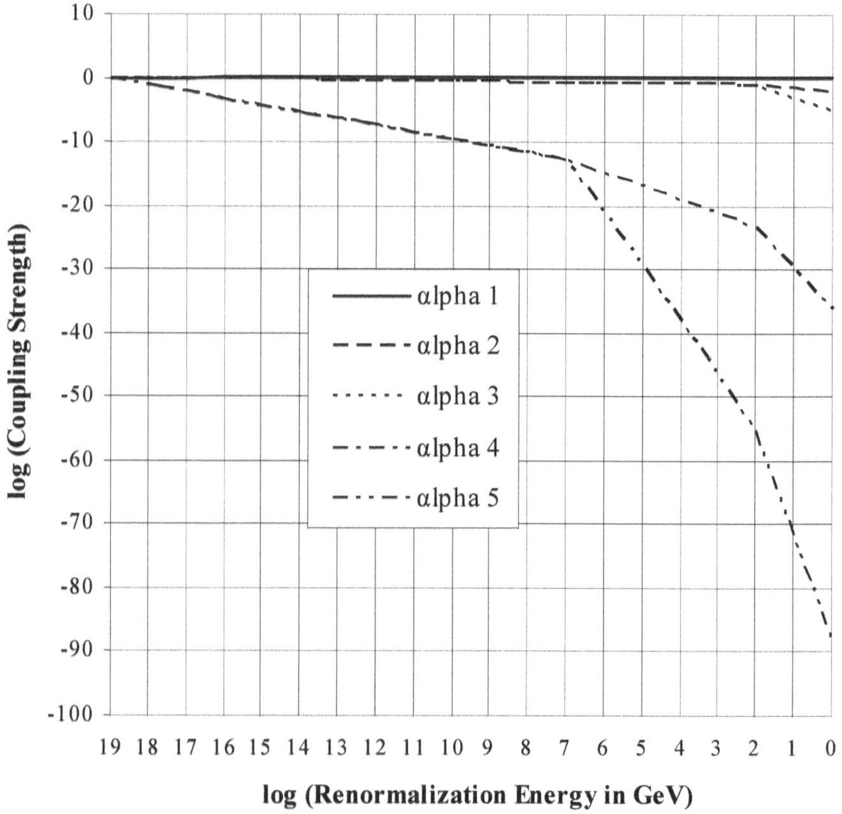


Figure 1 reviews the evolution (via the RGE's, Equation (14) and this Section's assumptions) of the couplings down from a GUT near the Planck energy scale (the left edge of Figure 1). At intermediate renormalization energy scales, there are partial GUT's at the Gravity scale of $\sim 1.9 \times 10^{7}$ GeV and the Electroweak scale of $\sim 90$ GeV. And finally, we have five distinct forces at low energy scales (the right edge of Figure 1).

Although it may be possible for more forces to exist, the sixth force is expected to be extremely weak $\left(\sim 10^{-193}\right)$, and this is much smaller than the inverse of the expected number of particles in the Universe ($> 10^{100}$ and probably no more than $\sim 10^{120}$ – Dirac's large number cubed in three dimensions, or the inverse of Einstein's Cosmological "Constant"), so the probability of interaction is practically zero. Furthermore, the higher $n$ GUM bosons and higher-level Clifford algebras seem to imply more massive, more charge-shielded bosons (such as the progression of π's, W's, Z's and WIMP-Gravitons). If sixthons were more massive than the Planck mass (Section 5.4 predicts $m^*_{6\ \mathrm{corr}} \sim 10^{46}\ \mathrm{GeV}/c^2 \sim 10^{27}\, m_{Planck} \sim 10^{19}$ kg $\sim 10^{-11}\, M_{Solar}$) then they may exceed the limitations of quantum spacetime and form some of Stephen Hawking's Mini Black Holes [24]. Thus, quantum four-dimensional spacetime and the Planck mass may demand a maximum of five forces.

If we connect this theory with String Theory, then larger $n$ GUM bosons may be higher-energy vibrations on a string (we had to make an ad-hoc assumption of quantized GUM boson momenta, which could easily be explained with String Theory). It would be interesting to pursue whether a string with at least seven dimensions (four spacetime dimensions plus at least the three string membrane dimensions from Section 4.2 – also, the $SO(6) \rightarrow SO(3) \times SO(3)$ of WIMP-Gravity in this Section may imply three extra dimensions) can be reduced, one dimension at a time with the appropriate action integrals, and extremized to give a scalar quartic-energy (perhaps energy-to-the-fourth because there are four observable dimensions – spacetime?). If so, then a solid connection between string theory, field theory and QSGUT may yet exist; and this book's modeling assumptions may yet be justified by more than their ability to fit the low-energy coupling constants for the electromagnetic, weak and gravitational forces.

## 5.6 Information Theory and QSGUT

Is the concept of a Grand Unified Theory such as QSGUT consistent with Information Theory? The argument is that a complicated GUT may introduce more complexity than the data it is trying to explain, and that an arbitrarily complex theory can explain anything (such as String Theory with $10^{500}$ parameters). Table 5 is a comparison of the unexplained data from the Standard Model (SM) on the left (all coupling constants are assumed measured in the low renormalization mass limit and denoted within QSGUT's framework) versus the new parameters from QSGUT on the right.

**Table 5 - A Comparison of Information Content**

| Section | SM: 7 Unknowns | QSGUT: 4-6 Unknowns |
|---|---|---|
| 4.2 | $\alpha_1, \alpha_2, \alpha_3, \alpha_4$ | $C_1, C_2, D?$ |
| 5.4 | $m_1, m_2, m_5$ | $C_3, C_4, \varepsilon?$ |

Have we reduced the number of parameters (and the amount of information) involved? We may have, considering that we may be able explain 1) $D = 3$ via String Theory, and 2) the value of $\varepsilon$ if it is related to $C_3$ or $\alpha_2$. Note that the chemical potential, $\mu^4$, is determined by the Thermodynamic Properties of QSGUT in Section 5.2, so this is not an unknown parameter. If future experiments can verify some of the predicted quantities, such as $\alpha_5$, $\partial\alpha_4/\partial\mu$, $m_3$ or $m_4$, then QSGUT will have phenomenological advantages over the Standard Model.

Have we violated the postulates of Algorithmic Information Theory? Algorithmic Information Theory views the Laws of Physics as a "Grand Computer Program". If a new theory adds too many new lines of computer code, then we should weigh (and question) its contribution to the field of Physics. This theory uses an existing theory, Quantum Thermodynamics, with two paradigm-shifts: 1) we used energy-to-the-fourth rather than energy for our probabilities – this concept was never disallowed by Thermodynamics, it simply never had a motivating application, and 2) we used Quantum Thermodynamics and Blackbody Radiation to gain a better understanding of the low-energy coupling "constants" and Grand Unification. The third new twist used the Correspondence Principle, which was a previously existing part of the Philosophy of Physics that originally demonstrated the similarities and differences between Classical and Quantum (or Relativistic) Mechanics, but now relates QSGUT with the Renormalization Group Equations of Quantum Field Theory and explains the "Dark Energy" problem. And the fourth new idea borrows the mass relation from the Semiclassical Model of Bloch electron dynamics in Solid State Physics, and reapplies a modified version of this concept to determine equivalent masses for the fundamental forces and gives a deeper understanding of the Higgs vev, Dark Matter and Inflation. Ultimately, these paradigm-shifts account for about four "new lines of computer code", but the contributions to a better understanding of Grand Unification, Dark Energy, Dark Matter, Inflation, the Higgs vev, and a possible reduction in the number of "random" Standard Model parameters, may justify the added algorithmic information.

# 6. Experimental Validations and Consequences

## 6.1 Current Experiments

Ultimately, a theory is useless without supporting data. This theory can fit measured coupling strengths, but can this theory predict anything new and measurable? Possible channels of endorsement include the cosmic ray spectrum, measured changes in the coupling "constants", the Chiao transducer, and the Podkletnov effect.

The cosmic ray spectrum [25] has a "knee" centered near a primary energy of $\sim 10^4$ TeV that increases the flux from $\sim 10^2$ TeV up to $\sim 10^6$ TeV. Although explanations consistent with the Standard Models of Particle Physics and Astrophysics have been proposed, new exotic particles at the $m_4^* c^2 \sim 1.9 \times 10^4$ TeV energy scale (the Gravity GUT scale with new bosons and fermions – see the next Chapter) could easily explain this knee.

Barrow and Webb [26] claim to have measured a variable electromagnetic coupling, $\alpha_2$, on the order of 5-10 parts per million weaker about 11-13 Billion Years ago. This timeline is similar to the apparent deceleration of stars about 10 Billion Years ago. According to Equation (15), such a decrease in $\alpha_2$ should be linked to a decrease in $\alpha_4$ of up to 2% (we have one generation of matter at low energies, so that $b_E = {}^{-32}\!/_9$), which causes a decrease in luminosity (a larger absolute magnitude) of 8% ($\exp(\text{Absolute Magnitude}) \sim (\text{Mass})^{-4} \sim \alpha_4^{-4}$), and causes these stars to appear to be 4% farther than they are ($(\text{Distance})^2 \sim \exp(\text{Apparent Magnitude} / \text{Absolute Magnitude})$ –

a 400 Million Light Year revision); and could help explain the apparent deceleration and successive acceleration of cosmological expansion commonly attributed to Dark Energy. These results imply that the Universe is about 400 Million Years younger than previously thought.

Other confirmations might include efforts by Chiao [27] and Podkletnov [28]. Raymond Chiao's transducer is designed to convert photons into gravitons by using a superconducting material, then to convert these gravitons back into photons, and use interferometer techniques to measure any changes. Chiao and his colleagues have not succeeded thus far, but they are using low-energy microwaves, and should consider that they may need to overcome or quantum-tunnel through a chemical potential of ~ 71 GeV (from Section 5.4). Although other scientific arguments expect a geometrical coupling of the electron's intrinsic spin with the graviton via spacetime's curvature, and particularly when superconductors are involved [29], Equations (8) and (14) – (15) specify the dynamics between photons and gravitons. The Podkletnov effect claims to produce a gravitational field via electromagnetic fields on a superconducting apparatus that is similar to a Van de Graaff generator, and is proportional to the mass it is acting on as a gravitational field should be, and seems to increase exponentially with electric potential up to the 2 MeV to which it has been tested. This increase in the effect is consistent with the possibility that this is quantum-tunneling from an apparatus operating far below its threshold energy (of 71 GeV). Is the Podkletnov effect an electric dipole effect, or are photons being converted into gravitons or WIMP-gravitons?

## 6.2 Variable Coupling Theory Revisited

Earlier, we discovered that Dirac's Large Number Hypothesis combined with the RGE's imply that old stars should burn too bright and the cosmological expansion is decelerating. This conclusion is wrong. In reality, stars less than ten billion light-years (LY's) away are too dim, whereas older and more distant quasars are anomalously bright. To further confuse the issue, the stars that are most unusually dim are about five billion LY's away. Many people try to explain this apparent deceleration and successive acceleration of cosmological expansion in terms of Einstein's Cosmological Constant, Dark Energy, Vacuum Energy, or a long-ranged repulsive Fifth Force [30].

However, there is a solution to this problem that requires no new and unusual physics other than this theory. The idea has three main components. First, the Dirac Large Numbers solution, $\alpha_4 = aT^{-1}$ was valid during the quasar-dominated era up until about ten billion years ago (If the Universe was about four billion years old at the time, then this is still a Large Number compared to typical electromagnetic transition times and Dirac's idea was not wrong, just misapplied). At this time, the Universe attempted to reach a minimum in GUM boson energy-mass dynamics (Section 5.4). However, the gravitational constant (and all of these coupling "constants", see Equations (25) and (26)) overshot its ideal value, which we assume would have been near the present value. These constants would have continually weakened (except the Strong coupling which strengthened) from ten until five billion years ago causing stars to be too dim because gravity was weaker than it is today. GUM boson energy-mass dynamics eventually reversed this trend about five billion years ago and brought the gravitational constant back up to its present value. Now we have a harmonic dependence imposed on the standard Hubble theory. This oscillation has a period of about twenty billion years, and is most consistent with standard Hubble theory if the oscillations are damped and asymptotically approaching an ideal value. Dirac expected Large Numbers to be proportional to within a factor of order unity, and (14 Billion Years)/(4 Billion Years) = 3.5 (the expected LNH evolution of the coupling large number since it diverged from the time large number 10 billion years ago) is a factor of order unity. If the gravitational constant varied in the distant past, but asymptotically approached a stable value in the past four billion years, then this variation might not affect General Relativity's interpretation of nearby astrophysical phenomena, and would leave few observable fossil consequences.

Thus, Variable Coupling Theory combines Dirac's Large Number Hypothesis first with Non-Equilibrium, and finally Equilibrium Energy-Mass Dynamics. Such a history of the gravitational "constant" would cause the Universe to *appear* to expand linearly for the most recent era, to *appear* to expand at an accelerating rate for stars about five billion LY's away, to *appear* to expand at a decelerating rate for stars about ten billion LY's away, and cause quasars to *appear* to be too bright. These perceptions would be true even if the underlying expansion was roughly linear, although the expansion may speed-up (slow-down) when gravity is weaker (stronger).

Figure 2 is a pictorial review of Variable Coupling Theory. This timeframe is based on the "Dark Energy" timelines of Reference [30] with the assumption of instantaneous communication of changes to all quantum states of the GUM boson and a 400 Million Year adjustment to the age of the Universe. The approximate magnitude of variations is based on Reference [26] and Equations (14) and (22). The left edge of the chart (Age less than 4 BY) is based on Dirac's Large Numbers Hypothesis and immediately follows the rough results of Figure 1. The right tail (Age of 10 to 11 BY) implies that the strong force may have been stronger than its current value, and this feature may help explain how the Oklo Natural

Nuclear Reactor began (due to a stronger $\alpha_1$), and why $\partial\alpha_2/\partial t$ has been so small since Oklo's formation [31].

**Figure 2 - Variable Coupling Theory (VCT)**

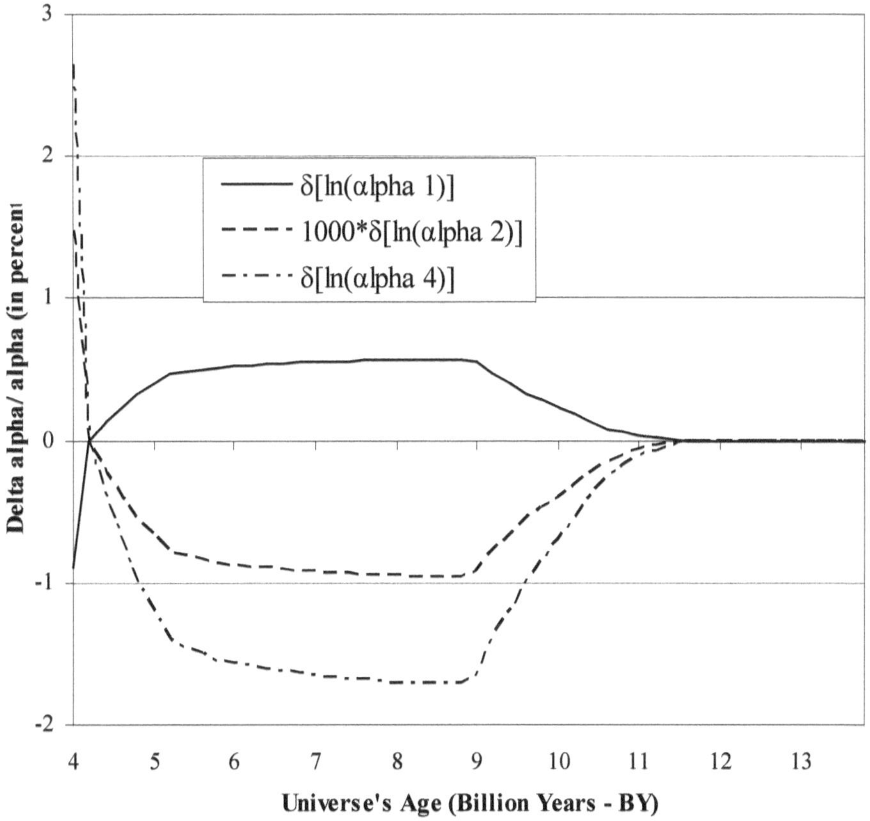

# 7. Speculations

Many scientists adhere to the paradigm of strong scientism – that the only "science" that deserves to be called Science is that which can be tested with the experimental method. This chapter's ideas fall outside of the realms of strong scientism; however, these topics seem relevant to GUT's, and future experiments may yet prove or disprove these ideas.

This chapter will use $SU(N)$, $SO(N)$ and $U(1)$ Lie Algebras (pronounced lē, not lī). A Special Unitary $SU(N)$ algebra is comprised of order $(N^2-1)$ matrices, each of rank $(N-1)$. A Special Orthogonal $SO(N)$ algebra is comprised of order $\frac{1}{2}N(N-1)$ matrices, each of rank $[truncate(N/2)]$. And the Unitary $U(1)$ algebra has order and rank of one.

## 7.1 A Grand Unification Lie Algebra Group

Within the Standard Model of Particle Physics, interactions of the Strong and Weak Nuclear, and Electromagnetic Forces (the spin-one gauge forces) are represented with the following product of Lie algebra groups [32] : $SU(3)_C \times SU(2)_L \times U(1)_Y$, each having a rank of two, one and one, respectively; and each having an order of eight, three and one, respectively. If we want to unify this spin-one gauge boson content within one Lie algebra group, we will need a group with the following requirements: 1) a minimum rank of four (2 + 1 + 1 = 4), 2) an order divisible by three (three gauge forces), and 3) an order large enough to contain the largest sub-order (eight gluons) after being divided by three. The smallest Lie algebra group that meets these conditions is *SU*(5) with a rank of four and an order of twenty-four (24 = 3 × 8), and introduces $X^{\pm 5/3}$ and $Y^{\mp 1/3}$ gauge bosons that can transform one generation of leptons into quarks and vice versa, thus independently violating lepton and baryon number conservation.

Another important Lie algebra unification group is *SO*(10), which has a rank of five, an order of forty-five (45 = 3 × 15), and contains three generations of fifteen spin-one-half matter fermions: $\left\{\left(u_{R,G,B}, d_{R,G,B}\right)_L, \left(e, \nu_e\right)_L, u_{R,G,B\,R}, d_{R,G,B\,R}, e_R\right\}$, $\left\{\left(c_{R,G,B}, s_{R,G,B}\right)_L, \left(\mu, \nu_e\right)_L, c_{R,G,B\,R}, s_{R,G,B\,R}, \mu_R\right\}$, and $\left\{\left(t_{R,G,B}, b_{R,G,B}\right)_L, \left(\tau, \nu_\tau\right)_L, t_{R,G,B\,R}, b_{R,G,B\,R}, \tau_R\right\}$ (two quark flavors times three colors times two spin polarizations plus two charged lepton spin polarizations plus one left-handed neutrino, 2 × 3 × 2 + 2 + 1 = 15) per generation, although the right-handed neutrino is an allowed singlet within this algebra group – in agreement with recent discoveries of neutrino mass oscillations [33].

To include all boson symmetry groups (if it is reasonable to include spin-one gauge and spin-two tensor bosons in one GUT Lie algebra), we now need to include the interaction Lie algebra groups for the spin-two interactions: Gravitation and WIMP-Gravitation, which, after symmetry-breaking, are expected to be $U(1)_G \times SU(4)_{WG}$. Now we have a product of five algebra groups: $SU(3)_C \times SU(2)_L \times U(1)_Y \times U(1)_G \times SU(4)_{WG}$; each having a rank of two, one, one, one and three, respectively; and each having an order of eight, three, one, one and fifteen, respectively. So that we may unify this boson content, we want a Lie algebra group with a minimum rank of eight (2 + 1 + 1 + 1 + 3 = 8), its order should be divisible by five (five forces), and it should still contain the largest sub-order (fifteen WIMP-Gravitons) after being divided. An obvious candidate is

$SU(9)$, with a rank of eight and an order of eighty ($80 = 5 \times 16$).

If we now assume that Supersymmetry (SUSY) – a symmetry that interrelates boson and fermion fields and naturally arises from String/ Supergravity Theory – is a true representation of nature, then we expect the superpartners of the spin-one gauge bosons to be spin-one-half gaugino fermions, and a full counting of spin-one-half fermions would contain at least sixteen matter fermions (the above fifteen fermions per generation plus the right-handed neutrino) and eight gluinos (our largest gaugino sub-order) for a total of twenty-four spin-one-half fermions in the smallest possible sub-order of the Grand Unification Lie algebra group. From the arguments of the previous paragraph, we still expect a GUT Lie algebra group that is divisible by five and has a minimum rank of eight. The smallest Lie algebra group that meets these conditions is $SU(11)$ with a rank of ten and an order of 120 ($120 = 5 \times 24$).

In a supersymmetric world, the above fermionic $SU(11)$ algebra with five generations should have a bosonic equivalent. Note that $SO(16)$ and $SU(11)$ both have an order of 120, and both add six new "charges" to the respective fermion and boson GUT groups of $SO(10)$ and $SU(5)$. In 1979, Howard Georgi proposed an $SU(11)$ GUT that could spontaneously break into an $SU(6)_{Flavor} \times SU(5)_{GUT}$ where the $SU(6)$ is related to flavor and the $SU(5)$ is our gauge boson GUT [34], but this ignores Gravity (which may be OK if we have another GUT group for tensor bosons). Suppose the bosonic $SU(11)$ algebra with five generations spontaneously breaks into an $SU(3)_C \times U(1)_Y \times SU(2)_L \times SU(3)_{HF} \times U(1)_G \times SU(4)_{WG}$ where the Electroweak force has a richer (and more massive) $U(1)_Y \times SU(2)_L \times SU(3)_{HF}$ gauge boson spectrum that the author calls Hyperflavor-Electroweak (HEW – see the next Section). Either Georgi's proposal or QSGUT's mass scales united with HEW may explain phenomenological questions as to why there are three low-mass generations of fermions; neutrino oscillations; and the Cabibbo-Kobayashi-Maskawa (CKM) Matrix that relates down, strange and bottom quarks [35].

If we reexamine the structure of Clifford algebra from Table 4, we realize that these are just combinatorial results, i.e. for the diagonal just inside the polar vectors, $N = n!/[2!\ (n-2)!]$ where $N$ = Order and $n$ = Clifford level $\geq 4$ (these are the $SO(n)$ Lie algebras with rank equal to $n/2$ for even $n$, or $(n-1)/2$ for odd $n$), and similarly for the next inner diagonal, $N = n!/[3!\ (n-3)!]$ for $n \geq 6$, and so on. Further study of these representations yields a triangle divisible by five (underlined in Table 4) that reaches from Clifford levels five to eight. This triangle may imply the five forces used here. Other large triangles also exist, such as a triangle divisible by seven (or fourteen) that covers Clifford levels seven (or eight) through twelve. Any $SU(14\,n \pm 1)$ Lie algebra (where $n$ is a definite positive integer) contains 14-plets, but $SU(29)$ is special because it is also divisible by the 120 of $SU(11)$ and the 28 of $SO(8)$. Many of the Lie algebra groups discussed thus far have corresponding Clifford algebras. $SU(29)$ is not one of these algebra groups, but there is a related 210-plet of $SO(21)$. Note that $SU(11)$ has an order of 120 = 5!/1!, whereas $SU(29)$ has an order of 840 = 7!/3! = 120 × 7 = 28 × 30 = 14 × 60. $SU(29)$ may be a grander unification including more forces, but this is phenomenological overkill because these extra forces are extremely weak with extremely heavy mediating bosons.

Are there higher levels of Grand Unification? The ultimate GUT group may be similar to an infinite onion with a new layer for every new prime number multiple. The next large triangle begins at Clifford level eleven, and is divisible by eleven.

## 7.2 The Hyperflavor-Electroweak Force

The concept behind Hyperflavor-Electroweak (HEW) is that the Electroweak force is actually a $U(1)_Y \times SU(2)_L \times SU(3)_{HF}$ Lie algebra, but eight $W^{\pm}_{2-4}$ and $B^0_{1,2}$ of its 12 IVB gauge bosons are so massive that they have not yet been discovered and, for all practical purposes, the Hyperflavor-Electroweak force may be represented as a broken $U(1)_Y \times SU(2)_L$ algebra. These new Hyperflavor bosons are expected to couple to the QSGUT gravity mass scale of $m^*_{4\ \mathrm{corr}} \sim 1.9 \times 10^4\ \mathrm{TeV}/c^2$. These $U(1)_Y \times SU(2)_L \times SU(3)_{HF}$ bosons could be the intra-generational generators needed to transform one generation of quarks $(u_L, d_L, u_R, d_R)$ or leptons $(e_L, \nu_{eL}, e_R, \nu_{eR})$ into each other. Extra-generational generators needed to transform one generation into another, such as within $(e, \mu, \tau)$, $(\nu_e, \nu_\mu, \nu_\tau)$, $(u, c, t)$ or $(d, s, b)$, may be hidden in the broken remnants of an *SU*(11) GUT (see Section 7.4). Table 6 reviews a possible configuration of HEW quantum numbers for the first generation of leptons (also see Figure 3) and quarks where all charges have been chosen so that the sum of charges within each generation equals zero. We can obtain the charges of a broken $(\mathrm{T}_{3L}, \mathrm{T}_{3R})$ (a Minimal Left-Right Symmetric Weak Model) and unbroken $(\mathrm{T}_{3L}, \mathrm{T}_{3HF}, \mathrm{T}_{8HF})$ HEW Theory in terms of the original left-handed weak isospin, $\mathrm{T}_{3L}$, the original left-handed weak hypercharge, $Y_L$, two new hyperflavor-weak isospins, $\mathrm{T}_{3HF}$ and $\mathrm{T}_{8HF}$, and a new right-handed weak hypercharge, $Y_R$, as follows:

$$\mathrm{T}_{3R} = \left(\sqrt{3}\ \mathrm{T}_{3HF} + \sqrt{6}\ \mathrm{T}_{8HF}\right)/3 \text{ and}$$
$$Q = \mathrm{T}_{3L} + \tfrac{1}{2} Y_L = \mathrm{T}_{3R} + \tfrac{1}{2} Y_R. \qquad (33)$$

**Table 6 - Hyperflavor-Electroweak Quantum Numbers**

| Particle | $T_{3L}$ | $\sqrt{3}\,T_{3HF}$ | $\sqrt{6}\,T_{8HF}$ | $Y_L$ | $Y_R$ | $Q$ | $T_{3R}$ |
|---|---|---|---|---|---|---|---|
| $e_L$ | $-1/2$ | $1/2$ | $-1/2$ | $-1$ | $-2$ | $-1$ | 0 |
| $\nu_{eL}$ | $1/2$ | $1/2$ | $-1/2$ | $-1$ | 0 | 0 | 0 |
| $e_R$ | 0 | $-1$ | $-1/2$ | $-2$ | $-1$ | $-1$ | $-1/2$ |
| $\nu_{eR}$ | 0 | 0 | $3/2$ | 0 | $-1$ | 0 | $1/2$ |
| $u_L$ | $1/2$ | $-1/2$ | $1/2$ | $1/3$ | $4/3$ | $2/3$ | 0 |
| $d_L$ | $-1/2$ | $-1/2$ | $1/2$ | $1/3$ | $-2/3$ | $-1/3$ | 0 |
| $u_R$ | 0 | 1 | $1/2$ | $4/3$ | $1/3$ | $2/3$ | $1/2$ |
| $d_R$ | 0 | 0 | $-3/2$ | $-2/3$ | $1/3$ | $-1/3$ | $-1/2$ |

**Figure 3 – A Tetrahedron of Leptons and HEW Bosons**

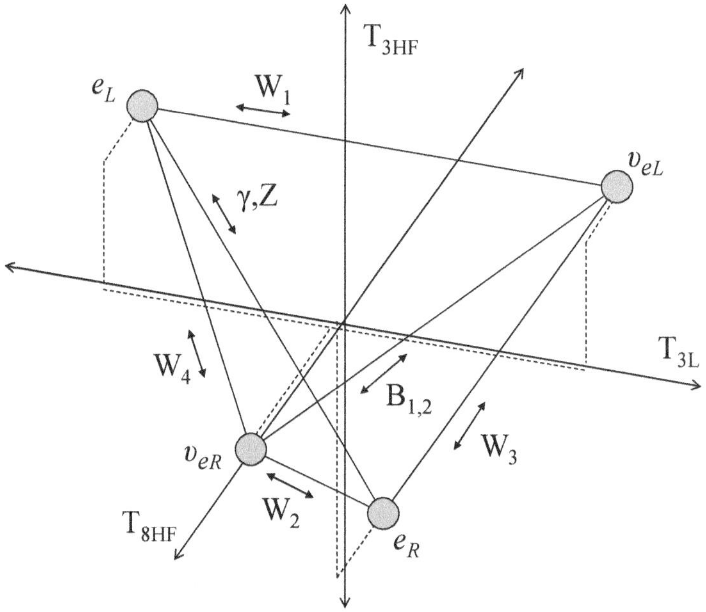

These leptons and quarks each map out the equiangular, equidistant vertices of two nested tetrahedra (also known as a stella octangula – a tetrahedron, or 3-simplex, has four equilateral triangular sides) in this 3-dimensional hyperflavor-weak "isospin charge space". Should we connect these particles via rotational point or translational space symmetries? The tetrahedral group, $\boldsymbol{T}_d$, has point group symmetries of order 24 in five classes: a singlet *1*, a 3-plet $\mathbf{C}_2$, an 8-plet $\mathbf{C}_3$, and two 6-plets $\mathbf{S}_4$ and $\sigma_d$ [36]. Coincidentally, this is how the *SU*(5) GUT breaks into $SU(3)_C \times U(1)_Y \times SU(2)_L$ plus a 6-plet (two electric times three color charges) each of $X^{\pm 5/3}$ and $Y^{\mp 1/3}$ leptoquark bosons. Nature already uses these symmetry groups for crystalline structures. Could these concepts also apply to a hyperflavor-weak isospin crystal? Section 5.4 used a mass relation from Solid State Physics without a crystal lattice, but hyperspace may be partially composed of hyperflavor-weak isospin 3-simplices that permit this mass analysis.

The tetrahedral group has a pure rotational subgroup, ***T***, of order 12 (similar to $U(1) \times SU(2) \times SU(3)$) in four classes: a singlet *1*, a 3-plet $\mathbf{C}_2$, and two 4-plets $\mathbf{C}_3$ and $\mathbf{C}_3^2$. On a grander scale, this stella octangula of quarks and leptons could be treated as a single cube (see Figure 4 on the front cover) of identical fermions, thus requiring the $X^{\pm 5/3}$ and $Y^{\mp 1/3}$ leptoquark bosons. The cube belongs to the octahedral group, $\boldsymbol{O}_h$, which has point group symmetries of order 48 (similar to *SU*(7)) in ten classes: two singlets *1* and *i*; two 3-plets $\mathbf{C}_2 = \mathbf{C}_4^2$ and $\sigma_h$; four 6-plets $\mathbf{C}_2, \mathbf{C}_4, \mathbf{S}_4$ and $\sigma_4$; and two 8-plets $\mathbf{C}_3$ and $\mathbf{S}_6$. The octahedral group has a pure rotational subgroup, ***O***, of order 24 in five classes that shares similarities with $\boldsymbol{T}_d$: a singlet *1*, a 3-plet $\mathbf{C}_2$, an 8-plet $\mathbf{C}_3$, and two 6-plets $\mathbf{C}_4$ and $\mathbf{C}_2$.

If we start with a 12-plet of gauge bosons and four complex scalar doublet fields, these twelve Hyperflavor-Electroweak bosons could spontaneously break into an $U(1)_Y \times SU(2)_L \times SU(3)_{HF}$ (similar to the two 4-plets, the 3-plet and the singlet of ***T*** – the tetrahedral rotational subgroup). The $SU(3)_{HF}$ and two complex scalar doublets could mix to produce supermassive $B^0_{1,2}$ and $W^{\pm}_{2-4}$ bosons and their longitudinal degrees of freedom, $\Phi_{B_{1,2}}, \Phi_{W_{2-4}}$ and $\Phi^*_{W_{2-4}}$. These two complex scalar fields give mass to the particles on the QSGUT supermassive scale of $m^*_{4eff} \sim 1.9 \times 10^4$ TeV / $c^2$. The remaining $U(1)_Y \times SU(2)_L$ unites with the 2 remaining complex scalar doublets to create the MSSM spectrum of massive $H^{\pm}, H_P, H_H, H_L$, $W_1^{\pm}$, and $Z^0$ bosons, and a massless photon $\gamma$. The degrees of freedom for this HEW-Higgs sector is 28, like an *SO*(8) algebra.

If we use translational symmetries, we need six translations and their inverses to connect all leptons (or quarks). If we define HEW isospin change as $\Delta\mathbf{T}_{mn} = \left(T_{3L,m} - T_{3L,n}\right)\mathbf{i} + \left(T_{3HF,m} - T_{3HF,n}\right)\mathbf{j} + \left(T_{8HF,m} - T_{8HF,n}\right)\mathbf{k}$, $\left|\Delta\mathbf{T}_{mn}\right| = \left(\Delta\mathbf{T}_{mn} \cdot \Delta\mathbf{T}_{mn}\right)^{1/2}$ and $\mathbf{T}' = T_{3L}\,\mathbf{i} + \sqrt{3}\,T_{3HF}\,\mathbf{j} + \sqrt{6}\,T_{8HF}\,\mathbf{k}$, then we find that all transitions between different leptons $\left(m,n = e_L, \nu_{eL}, e_R, \nu_{eR}\right)$ or between different quarks $\left(m,n = u_L, d_L, u_R, d_R\right)$ have $\left|\Delta\mathbf{T}\right| = 1$. These transitions are mediated by the $W^{\pm}_{1-4}, B^0_{1,2}$ and $Z^0$ Hyperflavor-Weak Intermediate Vector Bosons or the photon. Four lepton-quark transitions $\left(e_L \leftrightarrow u_L,\ \nu_{eL} \leftrightarrow d_L,\ e_R \leftrightarrow u_R,\ \text{and } \nu_{eR} \leftrightarrow d_R\right)$ have $\left|\Delta\mathbf{T}\right| = \sqrt{3/2}$. All other lepton-quark and quark-lepton transitions have $\left|\Delta\mathbf{T}\right| = \sqrt{1/2}$. These transitions are mediated by the $X^{\pm 5/3}$ and $Y^{\mp 1/3}$ leptoquark gauge bosons from *SU*(5) GUT's. Note that $\mathbf{T}'_{3HF} = \sqrt{3}\,\mathbf{T}_{3HF}$ and $\mathbf{T}'_{8HF} = \sqrt{6}\,\mathbf{T}_{8HF}$.

The fact that there are two different $|\Delta\mathbf{T}|$'s for lepton-quark transitions (this is model-independent in three hyperflavor-weak isospin dimensions) implies that there may be a richer fermion structure at higher energies. These two nested tetrahedra may be parts of two nested face-centered cubic (FCC) close-packing lattices [37] with sites for new fermions with primitive translation vectors of $\Delta\mathbf{T}'_{prim} = (1,0,0)$, $(\frac{1}{2},\text{-}\frac{3}{2},0)$ and $(0,1,2)$ (these vectors and their opposites are the $W_1^{\pm}$, $Z^0$ and $\gamma^0$, and $W_2^{\pm}$ IVB Bosons, respectively – see Figure 3), or we have the alternative set of primitive translation vectors: $\Delta\mathbf{T}'_{prim} = (\frac{1}{2},\frac{3}{2},0)$, $(\frac{1}{2},\frac{1}{2},-2)$ and $(\frac{1}{2},\text{-}\frac{1}{2},2)$ (these vectors and their opposites are the $W_3^{\pm}$, $B_{1,2}^0$ and $W_4^{\pm}$ IVB Bosons, respectively). Each lepton (or quark) would have twelve nearest-neighbor leptons (or quarks). Some of these fermions might become accessible at higher energies, when the heavier $W_{2-4}^{\pm}$ and $B_{1,2}^0$ bosons can form and interact. We should expect at least this tetragonal 4-plet plus each of their twelve nearest neighbors for a 28-plet of leptons (an $SO(8)$ Lie algebra group – see Table 7) with 24 new supermassive hyperflavor leptons, and another 28-plet of quarks (with 24 new supermassive hyperflavor quarks) times three possible colors for 84 total quark degrees of freedom.

Together, these two nested FCC lattices form a simple cubic (SC) lattice (see Figure 4 on the front cover) with a lattice spacing of $1/\sqrt{2}$ shorter than the original FCC lattices, and allow us to define a new cubic set of hyperflavor-weak isopspin vectors:

$$\mathbf{T}_Z = (\mathbf{T}_{3L} + \mathbf{T}_{3R})/\sqrt{2} = \left[\mathbf{T}_{3L} + \left(\sqrt{3}\,\mathbf{T}_{3HF} + \sqrt{6}\,\mathbf{T}_{8HF}\right)/3\right]/\sqrt{2}$$
$$\mathbf{T}_X = (\mathbf{T}_{3L} - \mathbf{T}_{3R})/\sqrt{2} = \left[\mathbf{T}_{3L} - \left(\sqrt{3}\,\mathbf{T}_{3HF} + \sqrt{6}\,\mathbf{T}_{8HF}\right)/3\right]/\sqrt{2}, \text{ and} \qquad (34)$$
$$\mathbf{T}_Y = \mathbf{T}_Z \times \mathbf{T}_X = \left(2\,\sqrt{3}\,\mathbf{T}_{3HF} - \sqrt{6}\,\mathbf{T}_{8HF}\right)/3\sqrt{2}.$$

**Table 7 – An *SO*(8) of First Generation HEW Leptons & Some Quarks**

| Particle | $\mathbf{T}_{3L}$ | $\mathbf{T}'_{3HF}$ | $\mathbf{T}'_{8HF}$ | Q | $\mathbf{T}_{3R}$ | Particle | $\mathbf{T}_{3L}$ | $\mathbf{T}'_{3HF}$ | $\mathbf{T}'_{8HF}$ | Q | $\mathbf{T}_{3R}$ |
|---|---|---|---|---|---|---|---|---|---|---|---|
| $e_{1L}$ | $-1/2$ | $1/2$ | $-1/2$ | $-1$ | 0 | $\nu_{e1L}$ | $1/2$ | $1/2$ | $-1/2$ | 0 | 0 |
| $e_{1R}$ | 0 | $-1$ | $-1/2$ | $-1$ | $-1/2$ | $\nu_{e1R}$ | 0 | 0 | $3/2$ | 0 | $1/2$ |
| $e_{2L}$ | $-1/2$ | $-3/2$ | $3/2$ | $-1$ | 0 | $\nu_{e2L}$ | $1/2$ | $-3/2$ | $3/2$ | 0 | 0 |
| $e_{2R}$ | 0 | 1 | $-5/2$ | $-1$ | $-1/2$ | $\nu_{e2R}$ | 0 | 2 | $-1/2$ | 0 | $1/2$ |
| $e_{3L}$ | $1/2$ | $-5/2$ | $-1/2$ | $-1$ | $-1$ | $\nu_{e3L}$ | $-1/2$ | $-1/2$ | $7/2$ | 0 | 1 |
| $e_{4L}$ | $1/2$ | $-1/2$ | $-5/2$ | $-1$ | $-1$ | $\nu_{e4L}$ | $-1/2$ | $3/2$ | $3/2$ | 0 | 1 |
| $e_{3R}$ | $-1$ | 0 | $3/2$ | $-1$ | $1/2$ | $\nu_{e3R}$ | 1 | $-1$ | $-1/2$ | 0 | $-1/2$ |
| $e_{4R}$ | $-1$ | 2 | $-1/2$ | $-1$ | $1/2$ | $\nu_{e4R}$ | 1 | 1 | $-5/2$ | 0 | $-1/2$ |
| $e_{5L}$ | $1/2$ | $-1/2$ | $7/2$ | $-1$ | 1 | $\nu_{e5L}$ | $-1/2$ | $-5/2$ | $-1/2$ | 0 | $-1$ |
| $e_{6L}$ | $1/2$ | $3/2$ | $3/2$ | $-1$ | 1 | $\nu_{e6L}$ | $-1/2$ | $-1/2$ | $-5/2$ | 0 | $-1$ |
| $e_{5R}$ | 1 | 0 | $3/2$ | $-1$ | $1/2$ | $\nu_{e5R}$ | $-1$ | $-1$ | $-1/2$ | 0 | $-1/2$ |
| $e_{6R}$ | 1 | 2 | $-1/2$ | $-1$ | $1/2$ | $\nu_{e6R}$ | $-1$ | 1 | $-5/2$ | 0 | $-1/2$ |
| $e_{7L}$ | $3/2$ | $1/2$ | $-1/2$ | $-1$ | 0 | $\nu_{e7L}$ | $-3/2$ | $1/2$ | $-1/2$ | 0 | 0 |
| $e_{7R}$ | 0 | 1 | $7/2$ | $-1$ | $3/2$ | $\nu_{e7R}$ | 0 | $-2$ | $-5/2$ | 0 | $-3/2$ |
| $u_{2L}$ | $1/2$ | $3/2$ | $-3/2$ | $2/3$ | 0 | $d_{2L}$ | $-1/2$ | $3/2$ | $-3/2$ | $-1/3$ | 0 |
| $u_{2R}$ | 0 | $-1$ | $5/2$ | $2/3$ | $1/2$ | $d_{2R}$ | 0 | $-2$ | $1/2$ | $-1/3$ | $-1/2$ |

Although Table 7 mostly emphasizes leptons, we may generate an *SO*(8) of quarks by recognizing that the HEW isospin inversion operator, $\mathbf{T} \rightarrow -\mathbf{T}$, maps the 4-plet of leptons $(e_L, \nu_{eL}, e_R, \nu_{eR})$ into the 4-plet of quarks $(u_L, d_L, u_R, d_R)$, and some of these quarks are displayed in this Table.

**Table 8 - The Vector Bosons of Hyperflavor-Electroweak Theory**

| IVB | $\Delta\mathbf{T}_{3L}$ | $\Delta\mathbf{T}'_{3HF}$ | $\Delta\mathbf{T}'_{8HF}$ | $e_{1L} \leftrightarrow$ | $\nu_{e1L} \leftrightarrow$ | $e_{1R} \leftrightarrow$ | $\nu_{e1R} \leftrightarrow$ |
|---|---|---|---|---|---|---|---|
| $W_1^{\pm}$ | $\pm 1$ | 0 | 0 | $\nu_{e1L}, \nu_{e7L}$ | $e_{1L}, e_{7L}$ | $\nu_{e3R}, \nu_{e5R}$ | $e_{3R}, e_{5R}$ |
| $W_2^{\pm}$ | 0 | $\pm 1$ | $\pm 2$ | $\nu_{e4L}, \nu_{e6L}$ | $e_{4L}, e_{6L}$ | $\nu_{e1R}, \nu_{e7R}$ | $e_{1R}, e_{7R}$ |
| $W_3^{\pm}$ | $\pm \frac{1}{2}$ | $\pm \frac{3}{2}$ | 0 | $\nu_{e2R}, \nu_{e5R}$ | $e_{1R}, e_{6R}$ | $\nu_{e1L}, \nu_{e5L}$ | $e_{2L}, e_{6L}$ |
| $W_4^{\pm}$ | $\pm \frac{1}{2}$ | $\mp \frac{1}{2}$ | $\pm 2$ | $\nu_{e1R}, \nu_{e6R}$ | $e_{2R}, e_{5R}$ | $\nu_{e2L}, \nu_{e6L}$ | $e_{1L}, e_{5L}$ |
| $\gamma, Z$ | $\pm \frac{1}{2}$ | $\mp \frac{3}{2}$ | 0 | $e_{1R}, e_{4R}$ | $\nu_{e2R}, \nu_{e3R}$ | $e_{1L}, e_{3L}$ | $\nu_{e2L}, \nu_{e4L}$ |
| $B_{1,2}^{0}$ | $\pm \frac{1}{2}$ | $\pm \frac{1}{2}$ | $\mp 2$ | $e_{2R}, e_{3R}$ | $\nu_{e1R}, \nu_{e4R}$ | $e_{2L}, e_{4L}$ | $\nu_{e1L}, \nu_{e3L}$ |

Thus, we expect the lepton and quark spectrum to increase by a factor of seven for every generation at higher energy scales, giving us $112 = 7 \times (15+1)$ degrees of freedom of leptons and quarks per generation. Although the original $U(1)_Y \times SU(2)_L \times SU(3)_{HF}$ HEW quantum numbers $(\mathrm{T}_{3L}, \mathrm{T}_{3HF}, \mathrm{T}_{8HF})$ are unique, some of these fermions have degenerate $SU(2)_L \times SU(2)_R$ Minimal Left-Right Symmetric Weak quantum numbers $(\mathrm{T}_{3L}, \mathrm{T}_{3R})$. At low energies, this model reproduces the Standard Model Weak interactions (see Table 8), and justifies the addition of right-handed neutrinos with non-zero $\mathbf{T}'$ charges. Extending Equations (17) and (18) to five forces:

$$(\delta \mathcal{G})_{\beta^4, \mathcal{Q}_1} = 0 \text{ with } \mathcal{G} = N \sum_{n=1}^{5} \left[ \mathcal{Q}_1 \left(n^4 - 1\right) \mathrm{p}_n + \beta^4 \mathrm{p}_n \ln \mathrm{p}_n \right] \tag{35}$$

may require the existence of supermassive quarks (*SO*(8) quarks, $q_{2-7}$ where $q$ = (*u*, *d*, *c*, *s*, *t*, *b*) and/ or 4th and 5th generation leptoquarks –

See Section 7.4) that carry charges for all five forces to balance this equation. Table 8 reviews the Hyperflavor Electroweak (HEW) IVB bosons and certain transitions in which each boson participates. The low-mass bosons, leptons and transitions are underlined, thus demonstrating that the HEW IVB bosons and the *SO*(8) fermions can condense down to yield the Electroweak IVB bosons, and the observed leptons and quarks with preferred doublets of Weak Isospin $\left(e_{1L}, \nu_{e1L}\right)$ and $\left(u_{1L}, d_{1L}\right)$, and Electric Charge $\left(e_{1L}, e_{1R}\right)$ and $\left(u_{1L}, u_{1R}\right)$. The right-handed neutrino, $\nu_{e1R}$, has neither electric charge nor (significant) mass as does $d_{1R}$, so there is not a low-mass boson operator to connect it to other low-mass fermions. This model may require mixing of the $\gamma, Z^0$ and $B^0_{1,2}$ bosons to explain the $d_{1R} \leftrightarrow d_{1L}$ spin-flip transition. Note that the HEW IVB bosons have electric charge, $Q = \sqrt{2}\,\Delta\mathbf{T}_Z$ (see Table 8 and Equation (34)).

If we examine the rotational point symmetries of the *SO*(8) of leptons (or quarks) in Table 7, we find that it breaks into a 4-plet and two 12-plets of respective radii $|\Delta\mathrm{T}| = \sqrt{3/8}$ (for $\left(e_1, \nu_{e1}\right)_{L,R}$), $\sqrt{11/8}$ (for $\left(e_{2-4}, \nu_{e2-4}\right)_{L,R}$) and $\sqrt{19/8}$ (for $\left(e_{5-7}, \nu_{e5-7}\right)_{L,R}$) about the origin, HEW isospin $\mathbf{T} = 0$. These 12-plets would each form the centers of the faces of a dodecahedron. The dodecahedron has 12 pentagonal faces and 20 vertices, and shares the same symmetry group with the icosahedron, which has 20 triangular faces and 20 vertices. The icosahedral group, $\boldsymbol{I}_h$, has point group symmetries of order 120 (similar to an $SU(11)$ algebra) in ten classes: two singlets ***1*** and ***i***; four 12-plets $\mathbf{C}_5, \mathbf{C}_5^2, \mathbf{S}_{10}$ and $\mathbf{S}_{10}^3$; two 15-plets $\mathbf{C}_2$ and $\sigma$; and two 20-plets $\mathbf{C}_3$ and $\mathbf{S}_6$. The icosahedral group has a pure rotational subgroup, ***I***, of order sixty in five classes: a singlet ***1***, two 12-plets

$\mathbf{C}_5$ and $\mathbf{C}_5^2$, a 15-plet $\mathbf{C}_2$, and a 20-plet $\mathbf{C}_3$. These component symmetries show how an *SU*(11) GUT of order 120 may decompose into an $SU(10)\times U(1)$, an $SU(9)\times U(1)$, an $SU(8)\times U(1)$, an $SU(7)_{GUT+}\times U(1)_G\times SU(4)_{WG}$ (where the *SU*(7) consists of a 20-plet, a 15-plet, a 12-plet and a singlet) or an $SU(5)_{GUT}\times U(1)_G\times SU(6)_{NEW}$ (where Georgi's anticipated *SU*(6) consists of a 20-plet and a 15-plet). The *SU*(10), *SU*(9) and *SU*(8) algebras do not represent gravity/ WIMP-gravity in an expected manner. The next level of decomposition includes $SU(7)_{GUT+}\to SU(5)_{GUT}\times SU(3)_{HF}$ or $SU(6)_{NEW}\to SU(4)_{WG}\times SU(3)_{HF}$, and finally $SU(5)_{GUT}\to SU(3)_C\times U(1)_Y\times SU(2)_L$. Combining the apparently icosahedral symmetries of *SU*(11) with the tetragonal symmetries of Hyperflavor (as demonstrated in Table 6) could lead to three generations of quark or lepton 4-plets in a dodecahedron 12-plet (such as $\left((e_1,\nu_{e1}),(\mu_1,\nu_{\mu 1}),(\tau_1,\nu_{\tau 1})\right)_{L,R}$ – Georgi's proposal reinterpreted within this framework), or five generations of quark or lepton 4-plets in an icosahedron 20-plet (such as $\left((e_1,\nu_{e1}),(\mu_1,\nu_{\mu 1}),(\tau_1,\nu_{\tau 1}),(\mathrm{a}_1,\psi_1),(\mathrm{v}_1,\omega_1)\right)_{L,R}$ – see Section 7.4). The next step towards an *SU*(11) GUT is to determine a Lagrangian that includes proper field, matter and symmetry-breaking terms.

The progression of n-simplices from equilateral triangles ($n=2$) to tetrahedra ($n=3$) to pentachora ($n=4$) to hexatera ($n=5$) leads to a progression of nearest-neighbor sites of 6, 12, 20 and 30; which implies a progression of GUT groups from $SU(5)$ to $SU(11)$ to $SU(19)$ to $SU(29)$.

Now a legitimate question might be "How many generations of matter exist?" All of the stable chemical elements are comprised of just one generation of spin-one-half matter fermions: up and down quarks and electrons. However various medium-to-high-energy cosmic ray, collider and supercollider experiments from 1936 to date have discovered two more generations of matter and many of their properties [38], and data on the decay-width of the $Z$ boson indicates that there are only three light-mass neutrino flavors (less than $45.6\ \mathrm{GeV}/c^2$) [39]. But Reference [33] implies that neutrinos have mass, and the likelihood of new neutrino masses larger than $45.6\ \mathrm{GeV}/c^2$ deserves consideration. Furthermore, we expect most of the particles associated with Gravitation and WIMP-Gravitation to be either massless, such as the graviton, or very massive, of order $m^*_{4eff} \sim 1.9 \times 10^4\ \mathrm{TeV}/c^2$, although new neutrinos (if they exist) and gravitinos may be much lighter. Nonetheless, the expectation is that there should be at least five generations of fermions (the same as the number of forces) and possibly an infinite number of generations, rather than three.

### 7.3 String Theory, GUT Dimensionality and *SU*(11)

String theory starts with ten or more dimensions that condense down into spacetime and many curled-up string dimensions [40]. To be consistent with a ten dimensional string, this theory needs one time dimension, three space dimensions, three important string/ m-brane dimensions (D=3 from our Preliminary Results in Table 2), and three less important (these dimensions do not contribute any measurable effects here) string/ m-brane dimensions. Thus, we have a hierarchy of significance for these three triplets (a hierarchy of three generations, each with an underlying $SO(3)$ or $SU(2)$ symmetry?) of space and string/ m-brane – or hyperspace – dimensions. Another old string model includes 26 dimensions (time plus five hierarchal string quintets of hyperspace dimensions?). These string triplets (or 3-branes) and string quintets (or 5-branes) may arise from the underlying structure of three relatively strong gauge forces (strong, electromagnetic and weak) plus two relatively weak tensor forces (gravity and WIMP-gravity), and the anticipated generational structure of three relatively lightweight fermion generations (masses less than $\sim 0.3\ \text{TeV}/c^2$) plus two relatively heavy fermion generations (masses of $\sim 1.9 \times 10^4\ \text{TeV}/c^2$).

There is also an interesting near coincidence in the number of expected dimensions, D, and the rank, ($N$−1), of our expected $SU(N)$ GUT's: D = 4, 10, 26; and ($N$−1) = 4, 10, 28. This implies a relationship between a GUT group and the number of dimensions in which it is represented. For example, an $SU(5)$ GUT should exist in four dimensions, and an $SU(11)$ GUT should exist in ten dimensions. Combining this idea with the previous paragraph implies that $SU(11)$ should decompose as $SU(11) \rightarrow SU(5)_{GUT} \times SU(3)_{HF} \times U(1)_G \times SU(4)_{WG}$. Here, the massless graviton is generated by $SU(11)$ symmetry-breaking and is weak in our four dimensions because its flux lines permeate all ten dimensions. The $SU(5)$ GUT contains the Strong, Weak, and Electromagnetic forces in four dimensions which fragment into three space plus one time dimensions – probably due to the breaking of $SU(5)_{GUT} \rightarrow SU(3)_C \times SU(2)_L \times U(1)_Y$ and the creation of the massless photon, which establishes the speed of light, and sets time apart from space. The $SU(4)$ of WIMP-Gravity primarily exists in the three least significant hyperspace dimensions and appears extremely weak to us because few flux lines escape out to our 4-spacetime dimensions. And the $SU(3)$ of Hyperflavor coexists with gravity in the three important hyperspace dimensions (thus suggesting that Hyperflavor IVB's have a mass comparable to the gravity mass scale of $\sim 1.9 \times 10^4\ \text{TeV}/c^2$, rather than the Electroweak scale of $\sim 0.1\ \text{TeV}/c^2$). An $SU(29)$ or $SU(27)$ GUT (both contain $SO(8)$ 28-plets, but $SU(27)$ has a rank of 26 similar to the number of string dimensions) might decompose into a ten-dimensional *SU*(11) GUT, two five-dimensional *SU*(6)'s or *SO*(11)'s (the two least significant 5-branes of the 26 hyperspace dimensions – which might also break into 3-branes and 2-branes), and at least three two-dimensional *SU*(3)'s or *SO*(5)'s (to convert 3-branes into 5-branes – these algebras might include SUSY's fermionic and bosonic operators on the $\sim 1.9 \times 10^4$ TeV / $c^2$ mass scale).

To simplify this scenario, we might think of an *SU*(5) GUT of the Strong, Weak and Electromagnetic forces existing in our four dimensions of spacetime, Hyperflavor is sequestered to the 2-brane (the original 3-brane fragments into a 2-brane and the 7th dimension) that spans the fifth through sixth dimensions – the Flavorbrane, WIMP-Gravity is sequestered to the 3-brane that spans the eighth through tenth dimensions – the Gravitybrane, and Gravity can communicate with all ten dimensions via the seventh dimension [41]. If this picture is reasonable, then we see the genius behind a five-dimensional Kaluza-Klein Unification of Electromagnetism and Gravity, and we might expect the following near-unifications to have phenomenological consequences: a 5-dimensional $SU(5)_{GUT} \times U(1)_G$, a 7-dimensional $SU(7)_{GUT+} \times U(1)_G$ (including Hyperflavor) or a 10-dimensional $SU(11)_{GUT}$ (including WIMP-Gravity). We see that rank-4, *SO*(8) 28-plets arise naturally in a 4-dimensional spacetime. We also realize that the "power sources" for Gravity, Hyperflavor and WIMP-Gravity are bound up in higher hyperspace dimensions that we cannot see, but affect us nonetheless. Perhaps mass is quantized in higher dimensions (at least within multiples of one of the mass hierarchy scales of Section 5.4), but this discrete quantum number is relayed to our four spacetime dimensions via a badly broken and possibly sequestered Higgs sector that scrambles the original quantum numbers such that mass seems arbitrary. Now we understand why the four-dimensional Standard Model seems broken and ugly, because it cannot exactly represent a five-dimensional *SO*(10) "Fermion GUT" or one of the aforementioned "Boson GUT's" with five or more dimensions, because the **T** hyperflavor-weak isospin $\left(T_{3L}, T_{3HF}, T_{8HF}\right)$ lattice and hyperspace dimensions never inflated as much as space.

How much did the hyperspace dimensions inflate? If we push Dirac's Large Numbers Hypothesis to the limit, we might expect $\left(10^{40}\right)^4 = 10^{160}$ different "sites" in a fully inflated four-dimensional spacetime. But the fifth through tenth dimensions are likely not fully inflated, or else we should be able to see or easily measure them. If we consider the three strongest forces only (of the basic *SU*(5) GUT), then the largest "Large Number" we can make is the ratio of the strong and weak nuclear couplings $\sim 10^{12-13}$. Both of these nuclear forces are very short-ranged and may play important roles in these under-inflated higher dimensions. It is an interesting coincidence that these hyperspace dimensions come in 3-branes and $\left(10^{12-13}\right)^3 \sim 10^{36-39} \sim$ Dirac's Large Number. This implies that each 3-brane of under-inflated hyperspace dimensions (the 5$^{th}$ through 7$^{th}$ and 8$^{th}$ through 10$^{th}$ dimensions) has the same number of sites as any one of our four spacetime dimensions. Now we have $\left(10^{40}\right)^6 = 10^{240}$ (the inverse-square of Einstein's Cosmological Constant) different possible position vector "sites" on a ten dimensional string. In Statistics theory, the statistical error in a sample of $N$ items is $\sqrt{N}$. The number of particles in the Universe is less than $10^{120}$, which is the square root of $10^{240}$. Is the Universe just a statistical fluctuation around nothingness? This seems consistent with the "Free Lunch" Hypothesis [42], that the four spacetime dimensions borrowed an energy $\Delta E$ from the other string dimensions, created the Big Bang (gravitational fields have negative energy, and were $\sim 10^{40}$ times stronger prior to inflation), and then repaid the energy in time $\Delta t \sim h/(2\pi \Delta E)$ consistent with Heisenberg's Uncertainty Principle, so that the Universe had a net energy of zero.

The Fourier transform of our space lattice gives the momentum lattice (the reciprocal lattice of an FCC lattice is a Body-Centered-Cubic (BCC) Brillouin zone in Solid State physics, or the reciprocal lattice of an SC lattice is another SC lattice), which also has $\sim 10^{240}$ possible momentum vectors. Now there are $10^{240} \times 10^{240} \sim 10^{480}$ (Dirac's Large Number raised to the $12^{th}$ power) possible space/ momentum combinations on our ten dimensional string landscape. Of these $10^{480}$ initial condition variations allowed by the String "Theory of Everything/ Anything", we have $10^{240}$ three-dimensional space-momentum permutations times $10^{80}$ time-energy permutations times $10^{160}$ possible "hidden variables" associated with the $5^{th}$ through $10^{th}$ string dimensions. We have $\sim 10^{40}$ position sites on the Gravitybrane (the $8^{th}$ through $10^{th}$ dimensions), and this may explain Dirac's Large Number and the extreme feebleness of Gravity.

String theory should have Kaluza-Klein particles that are identical to the particles we already know except that their masses should be much larger. If we equate our "Hyperflavor" electrons with first-order Kaluza-Klein electrons, then we obtain an estimate of the size of our Hyperspace dimensions from Heisenberg's Uncertainty Principle: $R \sim \Delta x \sim h/(2\pi \Delta p) \sim h/(2\pi m_4 c) = 1.0 \times 10^{-23}\,\text{m}$ $\left(\text{or } 10^{-8}\,\text{fm}\right)$ $= 6.4 \times 10^{11}\ l_{Planck}$. In contrast, the proton is about a femtometer (fm) across (which is a hundred million times wider than *R*). And a length of $\sim 10^{12}$ times larger than the Planck length ($l_{Planck}$) is consistent with our prior assumption that the ratio of the strong and weak couplings ($\sim 10^{12-13}$) determines the size of our under-inflated hyperspace dimensions.

## 7.4 A Particle Spectrum for Hyperflavor *SU*(11)

Note that the $SU(7)_{GUT+} \times U(1)_G$ should decompose into an $SU(5)_{GUT} \times SU(3)_{HF} \times U(1)_G$ with 16 off-diagonal components that separate into two 6-plets, a 3-plet and another singlet (from the symmetries of $\boldsymbol{O}_h$ and $\boldsymbol{T}_d$, $48 \rightarrow (2\times 8)+(4\times 6)+(2\times 3)+(2\times 1)$ and $24 \rightarrow (1\times 8)+(2\times 6)+(1\times 3)+(1\times 1)$). These 16 off-diagonal components might be the four complex Higgs scalar doublet fields, where the singlet is the light Higgs $(H_L)$ boson, the 3-plet are the longitudinal polarizations $(1\ \Phi_Z$ and $2\ \Phi_W's)$ for the $Z^0$ and $W_1^{\pm}$ bosons, and the two remnant 6-plets are the $H_H, H_P$ and $H^{\pm}$ Higgs bosons and the longitudinal polarizations $(6\ \Phi_W's$ and $2\ \Phi_B's)$ for the $W_{2-4}^{\pm}$ and $B_{1,2}^0$ bosons. Now the *SO*(8) of the HEW-Higgs sector has a rank and dimensionality of four with diagonal components $\gamma^0, Z^0, B_1^0$ and $B_2^0$ (see Table 9) mixtures of which correspond to the conserved quantum numbers $Q, T_{3L}, T_{3HF}$ and $T_{8HF}$. Comparing this Hyperflavor-Electroweak Theory to Physiological Taste (the "flavor" pun demands it), this theory has four charges: $Q, T_{3L}, T_{3HF}$ and $T_{8HF}$ versus four (or five) human flavors: bitter, salty, sour, sweet and (the less accepted) umami. This HEW-Higgs sector is badly broken, but these quantum numbers remain related via Equation (33). Note that these Higgs bosons may be sequestered to the Flavorbrane, which may equate this brane to the Weakbrane of String Theory terminology. The *SU*(11) remnants contain a pair of off-diagonal *SO*(8) 28-plets that could connect the "identical" members of an octet to each other. A possible octet is $(u_{cL}, d_{cL}, e_L, \nu_{eL}, u_{cR}, d_{cR}, e_R$ and $\nu_{eR})$ where "*c*" is one color, but these

**Table 9 - A Potential *SU*(11) Boson GUT Scenario**

| 10-D $SU(11)$ ⟷ | | | | | | | | | | |
|---|---|---|---|---|---|---|---|---|---|---|
| 7-D $SU(7) \times U(1)$ ⟷ | | | | | | | | | | |
| 4-D $SU(5)$ ⟷ | | | | | | | | | | |
| $I$ | $g_{GC}$ | $g_{BC}$ | $X_C^{-5/3}$ | $Y_C^{1/3}$ | $\Phi_{B_2}$ | $\Phi^*_{W_4}$ | $V_1^{0*}$ | $V_2^{0*}$ | $Q_{1a}^{-1/6}$ | $Q_{4a}^{-1/6}$ |
| $g_{RM}$ | $g_3$ | $g_{BM}$ | $X_M^{-5/3}$ | $Y_M^{1/3}$ | $H_P$ | $\Phi^*_{W_3}$ | $V_4^0$ | $V_3^{0*}$ | $Q_{2a}^{-1/6}$ | $Q_{5a}^{-1/6}$ |
| $g_{RY}$ | $g_{GY}$ | $g_8$ | $X_Y^{-5/3}$ | $Y_Y^{1/3}$ | $H^+$ | $\Phi^*_{W_2}$ | $V_5^0$ | $V_6^0$ | $Q_{3a}^{-1/6}$ | $Q_{6a}^{-1/6}$ |
| $X_R^{5/3}$ | $X_G^{5/3}$ | $X_B^{5/3}$ | $\gamma$ | $W_1^+$ | $\Phi_Z$ | $\Phi^*_{W_1}$ | $U_{1c}^{-1/3}$ | $U_{4c}^{-1/3}$ | $R_{1c}^{1/6}$ | $R_{4c}^{1/6}$ |
| $Y_R^{-1/3}$ | $Y_G^{-1/3}$ | $Y_B^{-1/3}$ | $W_1^-$ | $Z^0$ | $W_2^+$ | $W_3^+$ | $U_{2c}^{-1/3}$ | $U_{5c}^{-1/3}$ | $R_{2c}^{1/6}$ | $R_{5c}^{1/6}$ |
| $\Phi_{B_1}$ | $H_H$ | $H^-$ | $H_L$ | $W_2^-$ | $B_1^0$ | $W_4^+$ | $U_{3c}^{-1/3}$ | $U_{6c}^{-1/3}$ | $R_{3c}^{1/6}$ | $R_{6c}^{1/6}$ |
| $\Phi_{W_4}$ | $\Phi_{W_3}$ | $\Phi_{W_2}$ | $\Phi_{W_1}$ | $W_3^-$ | $W_4^-$ | $B_2^0$ | $U_{7a}^{1/3}$ | $U_{8a}^{1/3}$ | $U_{9a}^{1/3}$ | $U_{10}^{0*}$ |
| $V_1^0$ | $V_4^{0*}$ | $V_5^{0*}$ | $U_{1a}^{1/3}$ | $U_{2a}^{1/3}$ | $U_{3a}^{1/3}$ | $U_{7c}^{-1/3}$ | $G$ | $F_1$ | $F_4$ | $F_9$ |
| $V_2^0$ | $V_3^0$ | $V_6^{0*}$ | $U_{4a}^{1/3}$ | $U_{5a}^{1/3}$ | $U_{6a}^{1/3}$ | $U_{8c}^{-1/3}$ | $F_2$ | $F_3$ | $F_{10}$ | $F_6$ |
| $Q_{1c}^{1/6}$ | $Q_{2c}^{1/6}$ | $Q_{3c}^{1/6}$ | $R_{1a}^{-1/6}$ | $R_{2a}^{-1/6}$ | $R_{3a}^{-1/6}$ | $U_{9c}^{-1/3}$ | $F_5$ | $F_{11}$ | $F_8$ | $F_{13}$ |
| $Q_{4c}^{1/6}$ | $Q_{5c}^{1/6}$ | $Q_{6c}^{1/6}$ | $R_{4a}^{-1/6}$ | $R_{5a}^{-1/6}$ | $R_{6a}^{-1/6}$ | $U_{10}^0$ | $F_{12}$ | $F_7$ | $F_{14}$ | $F_{15}$ |

Table 9 Legend: $F$ = Fifthons = WIMP-Gravitons, $g$ = Gluons, $G$ = Graviton, $\gamma$ = Photon, $H$ & $\Phi$ = Higgs & Longitudinal Polarizations, $I$ = Identity, $Q$, $R$, $U$ & $V$ = Grand Bosons, $W_1$ and $Z$ = Weak IVB's, Other $W$'s and $B$'s = Hyperflavor Weak IVB's, $X$ and $Y$ = Leptoquark Bosons, $c$ = color (R,G,B), and $a$ = anti-color (C,M,Y)

**Table 10 – An 8-Dimensional CKM-PMNS-Munroe Mixing Matrix**

$$\begin{pmatrix} d'^{-1/3}_c & s'^{-1/3}_c & b'^{-1/3}_c & \nu'^0_e & \nu'^0_\mu & \nu'^0_\tau & \psi'^{-1/6}_a & \omega'^{-1/6}_a \end{pmatrix}$$

$$= \begin{pmatrix} D_1 & V_1^{0*} & V_2^{0*} & U_{1c}^{-1/3} & U_{4c}^{-1/3} & U_{7c}^{-1/3} & Q_{1a}^{-1/6} & Q_{4a}^{-1/6} \\ V_1^0 & D_2 & V_3^{0*} & U_{2c}^{-1/3} & U_{5c}^{-1/3} & U_{8c}^{-1/3} & Q_{2a}^{-1/6} & Q_{5a}^{-1/6} \\ V_2^0 & V_3^0 & D_3 & U_{3c}^{-1/3} & U_{6c}^{-1/3} & U_{9c}^{-1/3} & Q_{3a}^{-1/6} & Q_{6a}^{-1/6} \\ U_{1a}^{1/3} & U_{2a}^{1/3} & U_{3a}^{1/3} & D_4 & V_4^{0*} & V_5^{0*} & R_{1c}^{1/6} & R_{4c}^{1/6} \\ U_{4a}^{1/3} & U_{5a}^{1/3} & U_{6a}^{1/3} & V_4^0 & D_5 & V_6^{0*} & R_{2c}^{1/6} & R_{5c}^{1/6} \\ U_{7a}^{1/3} & U_{8a}^{1/3} & U_{9a}^{1/3} & V_5^0 & V_6^0 & D_6 & R_{3c}^{1/6} & R_{6c}^{1/6} \\ Q_{1c}^{1/6} & Q_{2c}^{1/6} & Q_{3c}^{1/6} & R_{1a}^{-1/6} & R_{2a}^{-1/6} & R_{3a}^{-1/6} & D_7 & U_{10}^{0*} \\ Q_{4c}^{1/6} & Q_{5c}^{1/6} & Q_{6c}^{1/6} & R_{4a}^{-1/6} & R_{5a}^{-1/6} & R_{6a}^{-1/6} & U_{10}^0 & D_8 \end{pmatrix} \begin{pmatrix} d_c^{-1/3} \\ s_c^{-1/3} \\ b_c^{-1/3} \\ \nu_e^0 \\ \nu_\mu^0 \\ \nu_\tau^0 \\ \psi_a^{-1/6} \\ \omega_a^{-1/6} \end{pmatrix}$$

fermions can already be transformed into each other in one or two operations via the $\gamma$, $B$, $W$, $X$, $Y$ and $Z$ bosons. If we propose that the 4th and 5th generations contain (audio, $\psi$) and (video, $\omega$) leptoquarks [43] (some literary license has been applied to Reference [43] itself), then another likely octet is $(d_c, s_c, b_c, \nu_e, \nu_\mu, \nu_\tau, \psi_a \text{ and } \omega_a)$. To fit into this octet, the 4th and 5th generations must not contain both down-type quarks and neutrinos.

The $SO(8) \times SO(8)^*$ 56-plet of Grand Bosons of the $SU(11)$ remnants may decompose as three 12-plets of $V^0, Q_c^{1/6}$ and $R_c^{1/6}$ bosons, and a 20-plet of $18\,U_c^{-1/3}$ and $2\,U^0$ bosons (see Tables 9 and 10, which are implied by the symmetries of $\boldsymbol{I}_h$ and $\boldsymbol{O}_h$, $120 \to (2 \times 20) + (2 \times 15) + (4 \times 12) + (2 \times 1)$), and could explain the CKM Matrix and neutrino oscillations (as represented by the Pontecorvo-Maki-Nakagawa-Sakata (PMNS) Matrix [44] – both effects are caused by $V$ bosons), and connect all five generations of fermions. These $Q$, $R$, $U$ and $V$ Grand Bosons may be sequestered on the Gravitybrane. Note that $S$ and $T$ are not used to denote any bosons here because they have been reserved for SUSY and hyperflavor-weak isospin, respectively. The 12-plet of $V$ bosons breaks down into two $SO(3) \times SO(3)^*$ 6-plets. One of these 6-plets is the CKM matrix, and the other is the PMNS Matrix. An $8 = 3 + 3 + 2$ dimensional $SO(8) \times SO(8)^*$ CKM-PMNS-Munroe (or Hyper-Pontecorvo, his 1959 neutrino paper predates the others) Mixing Matrix is presented in Table 10. Special orthogonal matrix conditions fix the values of the diagonal $D_{1-8}$ matrix elements in terms of the other 28 complex unknowns. This CKM-PMNS-Munroe Matrix could be components of an $N = 8$ Supergravity model that combines with 4-dimensional spacetime to produce a 12-dimensional SUSY GUT (see Section 7.6). These components may also break flavor symmetries to explain mass disparities such as between the down (~4-8 MeV/$c^2$) and bottom (~4000 MeV/$c^2$) quarks. At an intermediate scale, the 12-plet of $V$ bosons may mix with 18 of the 20 $U$ boson degrees of freedom (the $U_{10}^0$ boson connects the 4th and 5th generation leptoquarks) to form a $6 = 3 + 3$ dimensional $SO(6) \times SO(6)^*$ 30-plet that interrelates the sextet: $(d_c, s_c, b_c, \nu_e, \nu_\mu$ and $\nu_\tau)$. This is the simplest union of the CKM and PMNS Matrices.

What are the properties of our 4th and 5th generation Hyperflavor leptoquarks? One clue is the fact that the hyperflavor-weak isospin inversion operator maps leptons into quarks and vice versa. Thus, the origin, HEW isospin $\mathbf{T} = 0$, has properties that are halfway between leptons and quarks.

What strong nuclear color charge does the origin have? Each quark has strong charge represented by one of the 3 Young-Helmholtz colors (c): (Red, Green, Blue), and each lepton has no net color charge. Is no net color equivalent to white (W) or black (BL)? White (black) is the presence (absence) of all three colors, or the presence (absence) of a color and its anti-color (a): (Cyan, Magenta, Yellow). Thus, the sum of 2 different colors is an anti-color (i.e. BL + R + G = W − B = Y) and the addition table for cyclically different colors (i.e. BL + R + G + B + R + G + B + …) repeats as: black (0), color (1), anti-color (2), white (3), color (4 = 1), anti-color (5 = 2), white (6 = 3), etc. Similarly, the addition table for cyclically different anti-colors repeats as black (0), anti-color (1), color (2), white (3), anti-color (4 = 1), color (5 = 2), white (6 =3), etc. Assuming leptons have white strong color charge, the quantum state between a "colored" quark and a "white" lepton is an "anti-colored" leptoquark, and the strong color charge for the origin and other leptoquarks may be represented by an anti-color.

We might expect the origin to have two different kinds of quantum states. One state would be a *u-e* leptoquark (see Figure 4 on the front cover, note that a diagonal stretching from the $u_L$ to the $e_L$ crosses the origin) that we will call "a" (for "audio") with an electric charge of $+1/6$ (averaging our electric charge using the fact that there are three colors (or anti-colors) per quark (or leptoquark), we have $3 \times \frac{1}{6} = \left[\left(3 \times \frac{2}{3}\right) - 1\right]/2$ ). Another state would be a *d-v* leptoquark that we will call "$\psi$" with an electric charge of $-1/6$ (similarly, $3 \times \text{-}\frac{1}{6} = \left[\left(3 \times \text{-}\frac{1}{3}\right) + 0\right]/2$ ). This fourth generation $\psi$ leptoquark and its fifth generation cousin, the ω leptoquark, are involved in the CKM-PMNS-Munroe mixing matrix. Recognizing that we are working within a cubic symmetry (see Figure 4 on the front cover), we can easily envision a 7-plet of a and $\psi$ leptoquarks (the origin; plus one cubic unit to the right, left, front, back, up and down in the directions specified by Equations (34) – see Table 11). Earlier, we saw that our two FCC lattices of leptons and quarks collectively comprised an SC lattice. Adding our SC lattice of the origin and similar leptoquarks, we now have a collective BCC lattice of leptons, quarks and leptoquarks. Another interesting extention

**Table 11 – An *SO*(8) of Fourth Generation HEW Leptoquarks**

| Particle | $T_{3L}$ | $T'_{3HF}$ | $T'_{8HF}$ | $Y_L$ | $Y_R$ | Q | $T_{3R}$ |
|---|---|---|---|---|---|---|---|
| $(a,\psi)_{1\,L,R}$ | 0 | 0 | 0 | $\pm 1/3$ | $\pm 1/3$ | $\pm 1/6$ | 0 |
| $(a,\psi)_{2\,L,R}$ | $1/2$ | $-1/2$ | $-1$ | $-(2/3, 4/3)$ | $(4/3, 2/3)$ | $\pm 1/6$ | $-1/2$ |
| $(a,\psi)_{3\,L,R}$ | $-1/2$ | $1/2$ | 1 | $(4/3, 2/3)$ | $-(2/3, 4/3)$ | $1/6$ | $1/2$ |
| $(a,\psi)_{4\,L,R}$ | 0 | 1 | $-1$ | $\pm 1/3$ | $\pm 1/3$ | $\pm 1/6$ | 0 |
| $(a,\psi)_{5\,L,R}$ | 0 | $-1$ | 1 | $\pm 1/3$ | $\pm 1/3$ | $\pm 1/6$ | 0 |
| $(a,\psi)_{6\,L,R}$ | $1/2$ | $1/2$ | 1 | $-(2/3, 4/3)$ | $-(2/3, 4/3)$ | $\pm 1/6$ | $1/2$ |
| $(a,\psi)_{7\,L,R}$ | $-1/2$ | $-1/2$ | $-1$ | $(4/3, 2/3)$ | $(4/3, 2/3)$ | $\pm 1/6$ | $-1/2$ |

of this SC lattice of leptons and quarks would be a collective FCC lattice of leptons, quarks, leptoquarks and quark-neutrinos. Here, our leptoquarks are *u-v* and *d-e* mixtures with electric charges of $+1/3$ and $-1/3$ , respectively. And our quark-neutrinos are *u-d-e-v* mixtures with neutral electric charge. This scenario has ill-defined fermion spins, does not seem to work with the CKM-PMNS-Munroe mixing matrix, and is an interesting possibility that will not be further pursued.

What helicity does the origin have? The origin must have $\pm 1/2$ helicity due to its affiliation with leptons and quarks. Basic SUSY might expect supermassive 4th and 5th generation fermions (on the gravity unification mass scale) to have helicity of $\pm 3/2$ as do gravitinos (the SUSY partner to the spin-2 graviton). Now the origin could have helicity of $\pm 1/2$ and $\pm 3/2$ to connect it to leptons, quarks and gravitinos. Thus, even though the origin has Hyperflavor-Weak isospin quantum numbers of $\mathbf{T} = 0$ as reviewed in Table 11, it should have electric charge, color and possibly gravity quantum numbers (as implied by possible helicity values of $(\pm\frac{1}{2},\pm\frac{3}{2})$) – even before $SU(11)$ symmetry-breaking, and the mass scales of Section 5.4.

Now we have six general types of fundamental fermions: 1) electron-type leptons $(e,\mu,\tau)$ with seven Hyperflavor subtypes times two helicities each, electric charge of $-1$, and no net color charge (W), 2) neutrino-type leptons $(\nu_e,\nu_\mu,\nu_\tau)$ with seven Hyperflavor subtypes times two helicities each, no net electric charge (0), and no net color charge (W), 3) up-type quarks $(u,c,t)$ with seven Hyperflavor subtypes times two helicities, electric charge of $\frac{2}{3}$, and color charge of $(R,G,B)$ each, 4) down-type quarks $(d,s,b)$ with seven Hyperflavor subtypes times two helicities, electric charge of $-\frac{1}{3}$, and color charge of $(R,G,B)$ each, 5) audio-type leptoquarks $(\mathrm{a},\mathrm{v})$ with seven Hyperflavor subtypes times two helicities, electric charge of $+\frac{1}{6}$, and color charge of $(C,M,Y)$ each, and 6) psi-type leptoquarks $(\psi,\omega)$ with seven Hyperflavor subtypes times two helicities, electric charge of $-\frac{1}{6}$, and color charge of $(C,M,Y)$ each. This gives at least eighteen 28-plets of fermion degrees of freedom.

The flavor of the fermionic superstring depends on its location within this BCC hyperflavor-weak isospin lattice. Using Miller notation for the direct lattice in the $\mathbf{T}'$ cubic vectors of Equations (34) (also refer to Figure 4 on the front cover) [45], we have the following important body-diagonal directions: $\langle 111\rangle$ is parallel to some of the $Q$ and $R$ bosons in the $d_{1L} \leftrightarrow \psi_{1L} \leftrightarrow \nu_{e1L}$ transitions, $[1\bar{1}1]$ is important in the $e_{1L} \leftrightarrow \psi_{1L} \leftrightarrow u_{1L}$

transitions, $\left[11\bar{1}\right]$ is important in the $d_{1R} \leftrightarrow \psi_{1R} \leftrightarrow \nu_{e1R}$ transitions, and $\left[\bar{1}11\right]$ is important in the $e_{1R} \leftrightarrow \psi_{1R} \leftrightarrow u_{1R}$ transitions; the important face-diagonal directions: $\langle 101\rangle$, $\langle 10\bar{1}\rangle$, $\langle 011\rangle$ and $\langle 01\bar{1}\rangle$ are the $W^{\pm}_{1-4}$, respectively, and $\langle 1\bar{1}0\rangle$ and $\langle 110\rangle$ are the $\gamma$, $Z$ and $B_{1,2}$, respectively (see Table 8 for some of the transitions in which these bosons are involved); and the important edge directions: $\langle 100\rangle$ is important in the $u_{1R,L} \leftrightarrow \nu_{e1L,R}$, $e_{1L,R} \leftrightarrow d_{1R,L}$ and $\mathrm{a}_2 \leftrightarrow \psi_1 \leftrightarrow \mathrm{a}_3$ transitions, $\langle 010\rangle$ is important in the $u_{1L,R} \leftrightarrow \nu_{e1L,R}$, $e_{1L,R} \leftrightarrow d_{1L,R}$ and $\mathrm{a}_4 \leftrightarrow \psi_1 \leftrightarrow \mathrm{a}_5$ transitions, and $\langle 001\rangle$ represents some of the $X$ and $Y$ bosons in the $d_{1R,L} \leftrightarrow \nu_{e1L,R}$, $e_{1L,R} \leftrightarrow u_{1R,L}$ and $\mathrm{a}_6 \leftrightarrow \psi_1 \leftrightarrow \mathrm{a}_7$ transitions. Is one fermion more fundamental than any others? Opposing popular opinions that the electron is the most fundamental fermion, the psi-type leptoquark has hyperflavor-weak isospin $\mathbf{T} = 0$, is essential to several of these lepton-quark and intra-leptoquark transitions, and may qualify as the most fundamental fermion – despite its super-heavy mass (or maybe because it is defined to have mass!).

### 7.5 WIMP-Gravity Revisited

In Section 5.5 on WIMP-Gravity, we expected the *SU*(4) of WIMP-Gravity to have many pure imaginary components and collapse into a pseudo-*SU*(2) of real WIMP-Gravitons. Suppose the 15-plet of WIMP-Gravitons collapses into a pseudo-*SU*(2) 3-plet, the 56-plet of *Q, R, U* and *V* bosons collapses into the 30-plet of *U* and *V* bosons previously mentioned, and everything mixes to yield an effective *SU*(9) GUT that decomposes as: $SU(9) \rightarrow SU(7)_{GUT+} \times U(1)_G \times \left[\text{pseudo - }SU(2)_{3WG}\right]$.

In Table 9, the main diagonal components of WIMP-Gravity were numbered with traditional *SU*(4) nomenclature, i.e. 3, 8 and 15; however the cross diagonal components were numbered to respect the underlying *SO*(*N*) (with $N = 3$, 4, 5 or 6) symmetries. Analysis of these cross diagonals shows that these are particle/ anti-particle pairs. This allows us to rewrite the WIMP-Gravity *SU*(4) 15-plet from Table 9 as shown in Table 12. Note that there are 12 components of $F_A$ through $F_F$, thus these components may represent the pure rotational sub-group ***T*** of a new gravity-related tetrahedron. A suggested tetrahedron is shown in Table 13 (compare with Table 6), where the four vertices of the tetrahedron are matter, anti-matter, tachyonic matter, and tachyonic anti-matter. With these geometrical definitions, we may now define the off-diagonal WIMP-Gravitons, $F_A$ through $F_F$, as translations as shown in Table 14 (compare with Table 8).

**Table 12 – WIMP-Gravity Particle/ Anti-Particle Pairs**

$$\begin{pmatrix} G & F_1 & F_4 & F_9 \\ F_2 & F_3 & F_{10} & F_6 \\ F_5 & F_{11} & F_8 & F_{13} \\ F_{12} & F_7 & F_{14} & F_{15} \end{pmatrix} \rightarrow \begin{pmatrix} G & F_A^* & F_B^* & F_E^* \\ F_A & F_3 & F_D^* & F_C^* \\ F_B & F_D & F_8 & F_F^* \\ F_E & F_C & F_F & F_{15} \end{pmatrix}$$

This 15-pet of WIMP-Gravitons may spontaneously decompose into a Minimal Tardonic – Tachyonic Symmetric Gravity/ WIMP-Gravity Model:

$$U(1)_G \times SU(4)_{WG} \rightarrow \underbrace{U(1)_G \times SU(2)_{3WG}}_{\text{Tardonic Gravity}} \times \underbrace{U(1)_{8WG} \times SU(2)_{15WG}}_{\text{Tachyonic Gravity}} \tag{36}$$

where the $SU(2)_{3WG}$, $U(1)_{8WG}$ and $SU(2)_{15WG}$ consist of $\{F_A, F_A^*, F_3\}$, $\{F_8\}$, and $\{F_F, F_F^*, F_{15}\}$, respectively; $\mathrm{T}_{3i} = \left(\sqrt{3}\,\mathrm{T}_{8WG} + \sqrt{6}\,\mathrm{T}_{15WG}\right)/3 = \pm\tfrac{1}{2}$ defines tachyonic (imaginary) matter,

**Table 13 – A Tetrahedron of Real and Imaginary Matter**

| Type of Matter | $T_{3WG}$ | $\sqrt{3}\,T_{8WG}$ | $\sqrt{6}\,T_{15WG}$ | $T_{3i}$ |
|---|---|---|---|---|
| Matter $(+m)$ | $\frac{1}{2}$ | $-\frac{1}{2}$ | $\frac{1}{2}$ | 0 |
| Anti-Matter $(-m)$ | $-\frac{1}{2}$ | $-\frac{1}{2}$ | $\frac{1}{2}$ | 0 |
| Tacyonic Matter $(+im)$ | 0 | 1 | $\frac{1}{2}$ | $\frac{1}{2}$ |
| Tachyonic Anti-Matter $(-im)$ | 0 | 0 | $-\frac{3}{2}$ | $-\frac{1}{2}$ |

**Table 14 – The Off-Diagonal Tensor Bosons of WIMP-Gravity**

| WIMP-Graviton | $\Delta T_{3WG}$ | $\sqrt{3}\,\Delta T_{8WG}$ | $\sqrt{6}\,\Delta T_{15WG}$ | $\Delta T_{3i}$ |
|---|---|---|---|---|
| $F_A$ | $\pm 1$ | 0 | 0 | 0 |
| $F_B$ | $\pm \frac{1}{2}$ | $\mp \frac{3}{2}$ | 0 | $\mp \frac{1}{2}$ |
| $F_C$ | $\pm \frac{1}{2}$ | $\pm \frac{1}{2}$ | $\mp 2$ | $\mp \frac{1}{2}$ |
| $F_D$ | $\pm \frac{1}{2}$ | $\pm \frac{3}{2}$ | 0 | $\pm \frac{1}{2}$ |
| $F_E$ | $\pm \frac{1}{2}$ | $\mp \frac{1}{2}$ | $\pm 2$ | $\pm \frac{1}{2}$ |
| $F_F$ | 0 | $\pm 1$ | $\pm 2$ | $\pm 1$ |

and $T_{3WG} = \pm \frac{1}{2}$ defines tardonic (real) matter. The remaining 8-plet of WIMP-Gravitons allows interactions between tardons and tachyons.

Most of the WIMP-Gravitons are expected to have a mass of $m^*_{4\ \mathrm{corr}} \sim 1.9 \times 10^4\ \mathrm{TeV}/c^2$ (from Section 5.4). However, Section 5.4 also had a mass scale, $m^*_{3\ \mathrm{corr}} \sim i\ 55.8\ \mathrm{GeV}/c^2$, that had no apparent use (although this $SU(4)_{WG}$ of massive WIMP-Gravitons requires at least 16 more "Higgs-like" scalar degrees of freedom to provide longitudinal polarizations). Perhaps $F_{8WG}$ obtains the effective mass of $m^*_{3\text{ corr}} \sim i\,55.8\text{ GeV}/c^2$ and produces a stronger version of WIMP-Gravity than the off-diagonal components, $F_A$ through $F_F$. Section 5.5 estimated the strength and low-mass dependence of tardonic WIMP-Gravity as $1.34\times10^{-93}\times\left(m/m_{proton}\right)^{18.7}$. With these assumptions, tachyonic $U(1)_{8WG}$ WIMP-Gravity would have a coupling strength of $\frac{1}{3}\times1.34\times10^{-93}\times\left(55.8\text{ GeV}/c^2/0.938\text{ GeV}/c^2\right)^{18.7}\sim8\times10^{-61}$, where the factor of one-third is the ratio of degrees of freedom between one $U(1)_{8WG}$ and three $SU(2)_{3WG}$. This $U(1)_{8WG}$ coupling is the square root of Dirac's Large Number weaker than Gravity.

### 7.6 Beyond *SU*(11) – A Supersymmetric SUSY *SU*(15)?

What Lie algebras and string dimensions are required to represent Supersymmetry (SUSY)? Any further extensions to this GUT Lie algebra surpass our icosahedral model, and become more speculative. The simplest version of SUSY might require a 1-dimensional $SU(2)_{S\frac{1}{2}}$ with half-spin up and down "*S*" operators to relate the spin doublet (fermion, boson). More sophisticated algebras might include a 2-dimensional $SU(3)_{S1}$ with spin operators relating the spin triplet $(0,\frac{1}{2},1)$, or a 4-dimensional $SU(5)_{S2}$ with "Hyper-SUSY" spin operators that relate the spin quintet $(0,\frac{1}{2},1,\frac{3}{2},2)$.

Considering that the three most dominant 5-branes have fractured into 3-branes and 2-branes, we realize that a 10-dimensional non-SUSY $SU(11)$ GUT could combine with remnant 2-branes to become a 12- or 14-dimensional SUSY $SU(13)$ or $SU(15)$ GUT. Recent string theory developments by Edward Witten [46] prefer an 11- or 12-dimensional string. Eleven-dimensional models may include: $SU(7)_{GUT+} \times U(1)_G \times SU(5)_{S2}$ (without WIMP-Gravity), or $SU(11)_{GUT} \times SU(2)_{S1/2}$ (with a simplified SUSY). Twelve-dimensional models may include: $SU(7)_{GUT+} \times U(1)_G \times SU(2)_{WG} \times SU(5)_{S2}$ (with a simplified WIMP-Gravity), or $SU(13) \rightarrow SU(11)_{GUT} \times SU(3)_{S1}$, (with a simplified SUSY). But these SUSY GUT's should be interpreted as subsets of at least a 14-dimensional (one time, three hierarchal space/ string 3-branes, and the first two hierarchal 2-branes) SUSY $SU(15)$ GUT within the assumptions of this Section. Both $SU(13)$ and $SU(15)$ contain $SO(8)$ 28-plets which may explain the $SO(8)$ subsets of HEW-Higgs bosons; Grand bosons; and Hyperflavor Leptons, Quarks and Leptoquarks. The 24 Hyper-SUSY $SU(5)$ $S$ operators include four diagonal spin-0 scalar bosons and two off-diagonal $SO(5)$'s with six fermions (four spin-$\frac{1}{2}$ and two spin-$\frac{3}{2}$) and four bosons (three spin-1 and one spin-2) each as underlined in Table 15. Note that Table 15 is a diagonal extension of Table 9. This SUSY $SU(15)$ requires eight more $SO(5)$ groups of particles. SUSY allows fermions and their SUSY partners to be included in a boson GUT.

Standard SUSY nomenclature uses $\widetilde{G}$ (gravitino) as the spin-$\frac{3}{2}$ SUSY partner to the spin-2 $G$ (graviton), or $\widetilde{W}^{\pm}$ (wino) as the spin-$\frac{1}{2}$ SUSY partners to the spin-1 $W^{\pm}$, or $\widetilde{H}^{\pm}$ (higgsino) as the spin-$\frac{1}{2}$ SUSY partners to the spin-0 $H^{\pm}$ (charged higgs) – and linear combinations of these last two

**Table 15 - Detail of the SU(5) SUSY S Operators**

| | | | | | | |
|---|---|---|---|---|---|---|
| ↑ 14-D SUSY $SU(15)$ ↓ | ← 12-D $SU(13)$ → | $F_{15}$ | $\underline{S}_{1/2}$ | $\underline{S}_1$ | $\underline{S}_{3/2}$ | $\underline{S}_2$ |
| | | $S_{-1/2}$ | $S_0$ | $\underline{S}_{1/2}$ | $\underline{S}_1$ | $\underline{S}_{3/2}$ |
| | | $S_{-1}$ | $S_{-1/2}$ | $S_0$ | $\underline{S}_{1/2}$ | $\underline{S}_1$ |
| | | $S_{-3/2}$ | $S_{-1}$ | $S_{-1/2}$ | $S_0$ | $\underline{S}_{1/2}$ |
| | | $S_{-2}$ | $S_{-3/2}$ | $S_{-1}$ | $S_{-1/2}$ | $S_0$ |

SUSY states yield the $\widetilde{C}^{\pm}_{1,2}$ charginos. Although these symbols work with the MSSM where SUSY operations either increase or decrease spin by $1/2$, there is not such a simple solution with Hyper-SUSY spin operators that relate the spin quintet $(0, 1/2, 1, 3/2, 2)$. The author proposes a set of Hyper-SUSY operators: $\Sigma_{\pm 1/2}, \Sigma_{\pm 1}, \Sigma_{\pm 3/2}$ and $\Sigma_{\pm 2}$, such that $\widetilde{G} = \Sigma_{-1/2} G$, $\widetilde{H}^{+} = \Sigma_{1/2} H^{+}$ or $\widetilde{W}^{+} = \Sigma_{-1/2} W^{+}$. With this definition of Hyper-SUSY, we can connect spin-0 scalar bosons with spin-2 tensor bosons via the $\Sigma_{\pm 2}$ operators, or we can connect spin-$1/2$ fermions with spin-2 tensor bosons via the $\Sigma_{\pm 3/2}$ operators, etc. The prior paragraph needs eight $SO(5)$'s to fill in the off-diagonal elements of SUSY $SU(15)$. These $SO(5)$'s could be the eight $\Sigma_{\pm 1/2}, \Sigma_{\pm 1}, \Sigma_{\pm 3/2}$, and $\Sigma_{\pm 2}$ Hyper-SUSY operators applied to one of the $SO(5)$'s of the prior paragraph (see Table 16 at the end of this Section).

If these SUSY-related dimensions were formed by the union of two 2-branes, the SUSY-I-brane and SUSY-II-brane, then we should expect this

$SU(5)_{S2}$ to decompose into two 2-dimensional subgroups, such as $SU(3)_{S1} \times U(1)_{S0} \times SU(2)_{S½}$ with two off-diagonal $SO(4)$'s (see Table 15). This demonstrates the multifaceted nature of SUSY that leads to 11-, 12- or 14-dimensional SUSY GUT's via $SU(11)_{GUT} \times SU(2)_{S½}$, $SU(13) \rightarrow SU(11)_{GUT} \times SU(3)_{S1}$, or $SU(15) \rightarrow SU(11)_{GUT} \times SU(5)_{S2}$, respectively.

The similar SUSY $SU(15)$ fermion GUT should also contain our fermion matter particles. Note that $SU(15)$ has an order of $224 = 16 \times 14 = 8 \times 28 =$ eight $SO(8)$'s; such that this SUSY $SU(15)$ GUT could contain 28-plets of $(e_{1-7}, \nu_{e,1-7})_{L,R}$, $(u_{1-7,c}, d_{1-7,c})_{L,R}$, $(\mu_{1-7}, \nu_{\mu,1-7})_{L,R}$, $(c_{1-7,c}, s_{1-7,c})_{L,R}$, $(\tau_{1-7}, \nu_{\tau,1-7})_{L,R}$, $(t_{1-7,c}, b_{1-7,c})_{L,R}$, $(\mathrm{a}_{1-7,a}, \psi_{1-7,a})_{L,R}$ and $(\mathrm{v}_{1-7,a}, \omega_{1-7,a})_{L,R}$; not counting color or anti-color degrees of freedom. Another important hyperflavor sub-group is $SU(13)$, which has an order of 168, where $168 = 12 \times 14 = 6 \times 28 =$ six $SO(8)$ 28-plets, and would not include the fourth and fifth generation hyperflavor leptoquarks.

Extrapolating these ideas, the next level of GUT might require two more 2-branes, the Fermion-I-brane and Fermion-II-brane, that join to produce another diagonal $SU(5) \rightarrow SU(3) \times U(1) \times SU(2)$ with four more off-diagonal $SO(8)$'s in an 18-dimensional SUSY $SU(19)$ GUT (see Table 16). These $SO(8)$'s could contain one generation of fermions: $(e_{1-7}, u_{1-7,RGB})_{R,L}$ and $(\nu_{e1-7}, d_{1-7,RGB})_{R,L}$. This SUSY $SU(19)$ GUT (order of 360) contains many Lie algebra subgroups, but is also the smallest multiple of both the $SO(10)$ 45-plets and the $SU(5)$ 24-plets of the basic fermion and boson GUT's, which brings us 360 degrees "full circle" back to the roots of GUT. The anticipated 26-dimensional $SU(27)$ GUT (see Table 16) might contain this $SU(19)$ plus diagonal algebras: $SU(3) \times U(1) \times SU(6)$ with dimensionalities of 2 + 1 + 5, thus implying the union of the fourth 3-brane with the final 5-brane. This fourth 3-brane should fragment into an $SU(3)$ 2-brane (the Fermion-III-brane) plus a new $U(1)$ force. And this final 5-brane should fragment via $SU(6) \to SU(3) \times U(1) \times SU(3)$ into two $SU(3)$ 2-branes (the Fermion-IV-brane and Fermion-V-brane) plus a new $U(1)$ force. If the two off-diagonal 42-plets ($42 = SO(4) \times SO(7) \times SO(6)$) are scalars, then we have two "Higgs-like" polarizations for the diagonal $SU(3) \times U(1) \times SU(6)$ algebras except for the two massless, but very feeble $U(1)$ bosons.

We also have two off-diagonal $SO(16)$'s that each have the same order as $SU(11)$, and each could contain an 8-dimensional Fermion GUT. These $SO(16)$'s may fragment as $SO(16) \to SO(10) \times SO(7)$, where the ten charges of $SO(10)$ are 1) Quark, 2) Lepton, 3) Hypercharge, 4) Weak Isospin, 5-7) three Colors, and 8-10) three "Flavor" Generations; and the seven charges of $SO(7)$ are 1) Leptoquark, 2-3) two Hyperflavors, 4-5) two New Generations, 6) Basic SUSY, and 7) Electric Charge Conservation. The last charge of $SO(7)$ is due to $SU(5)$ GUT's, and this overlapping charge may cause the CKM-PMNS-Munroe Matrix. We have $U(1)$ symmetries that span the 13$^{th}$, 17$^{th}$, 21$^{st}$ and 24$^{th}$ dimensions and enforce the SUSY, Quark, Lepton and Leptoquark symmetries. And we have five Fermion-branes that lead to five generations of fermions.

## Table 16 – A 26-Dimensional $SU(27)$ GUT?

| | | | | | | |
|---|---|---|---|---|---|---|
| 0 | | | | | | |
| 1 | | | | | | |
| 2 | | | $SO(5)'$s of $\Sigma_{\frac{1}{2}}\{S*\}$ and | $SO(8)*$ | | |
| 3 | | | $\Sigma_{\frac{3}{2}}\{S*\}$ | | | |
| 4 | $SU(11)$ GUT w/ Spacetime, | | | | | |
| 5 | Weakbrane & Gravitybrane | | $SO(5)'$s of | | $SO(16)*$ has the same | |
| 6 | see Table 9 | | $\Sigma_1\{S*\}$ and | | order as $SU(11)$, | |
| 7 | | | $\Sigma_2\{S*\}$ | | but requires 8 | |
| 8 | | | | | dimensions, not 10 | |
| 9 | | | | | | |
| 10 | | | | $SO(8)*$ | | |
| 11 | $SO(5)$ of | $SO(5)$ of | $SU(5)$ | | | |
| 12 | $\Sigma_{-\frac{1}{2}}\{S\}$ | $\Sigma_{-1}\{S\}$ | SUSY- | | | |
| 13 | $SO(5)$ of | $SO(5)$ of | branes see | | | |
| 14 | $\Sigma_{-\frac{3}{2}}\{S\}$ | $\Sigma_{-2}\{S\}$ | Table 15 | | | |
| 15 | $SO(8)$ of | $SO(8)$ of | $SU(5)$ | $SO(4)$ | | |
| 16 | Hyperflavor | Hyperflavor | Fermion- | | | |
| 17 | Quarks? | Leptons? | branes? | | $SO(7) \times SO(6)$ | |
| 18 | | | | | | |
| 19 | | C | $SO(16) \rightarrow$ | $SO(4)$ | $SU(3)$ | |
| 20 | $SO(10)$ | K | $SO(10) \times SO(7)$ | | | |
| 21 | 3 Generations | M | 5 Gen's | | | |
| 22 | Fermion GUT? | P | Fermion GUT | | | $U(1) \times SU(6)$ |
| 23 | | M | | $SO(7) \times SO(6)$ | | Fermion- |
| 24 | | N | $SO(7)$ 4$^{th}$ & 5$^{th}$ | | | branes? |
| 25 | | S | Gen's? | | | |
| 26 | | M | | | | |

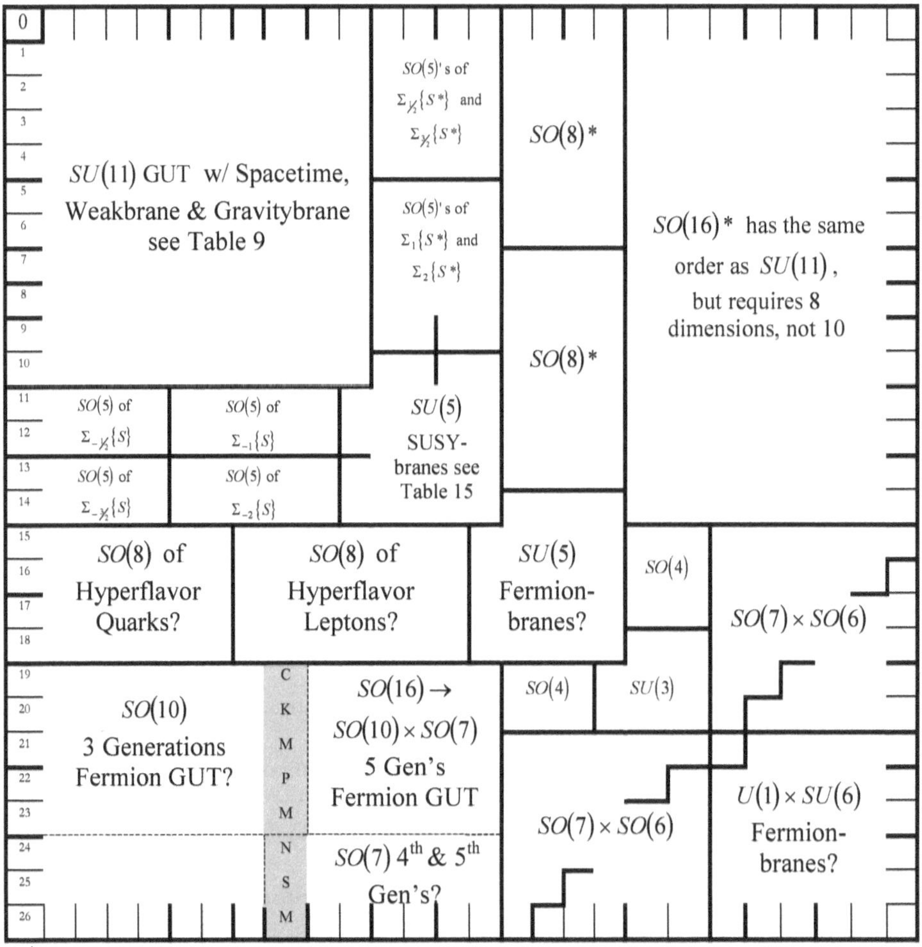

↑ Number of Dimensions

These $SO(16)'$s of Fermion GUT's have rank and effective dimensionality of eight, which implies compatibility with an $SU(9) \rightarrow SU(7)_{GUT+} \times U(1)_G \times \left[\text{pseudo-}SU(2)_{3WG}\right]$ Boson GUT group, and implies that the CKM-PMNS-Munroe Mixing Matrix elements are communicated via a broken Gravity/ WIMP-Gravity interaction along the seventh dimension.

## 7.7 A Deceptively Simple *E12* Theory of Everything (TOE)?

In 2007, A. Garrett Lisi posted "An Exceptionally Simple Theory of Everything" [47] in which he proposed a TOE based on the *E8* Lie Algebra. Although his theory is only 8-dimensional and does not include Hyperflavor or Five Generations of matter, it is an interesting and inspirational approach.

Is it possible to encompass all of the known bosons and fermions in one representation? The prior Sections represented fermions as a Body Centered Cubic (BCC) lattice in Hyperspace, and bosons as the Face Centered Cubic (FCC) lattice of vector operators that allow us to transition from one fermion site to another. Within this representation, fermions are the direct lattice in an extremely small Hyperspace lattice and bosons are the reciprocal lattice (the Brillouin zone) in an extremely large Universe superstructure. Lattice sites within a BCC (an FCC) lattice have 8 (12) nearest neighbors. If we assume that Minimal Supersymmetry (SUSY) is the Fourier transform that connects these direct and reciprocal lattices and allows these particles to coexist in one representation, then this leads to some interesting observations involving the Least Common Multipliers (LCM) and

**Table 17 – SUSY TOE Algebras for Select Lattice Combinations**

| # fermions | # bosons | LCM | Lie Algebra | Comments |
|---|---|---|---|---|
| 8 | 12 | 24 | $SU(5)$ | Basic BCC & FCC lattices |
| 16 | 12 | 48 | $SU(7)$ | Fermions with color |
| 56 | 12 | 168 | $SU(13)$ | Hyperflavor (HF) Fermions |
| 56 | 60 | 840 | $SU(29)$ | HF Fermions & 5 Forces |

Lie Algebra representations in Table 17. Note that the first three options contain three generations of fermions.

There is also an interesting near pattern in certain Exceptional Lie algebras, as shown in Table 18. Except for *E6*, this pattern is Order = $N = n \times [2 \times SO(n/2+2)+1] = (n^3 + 6n^2 + 12n)/4$, with a rank of *n*. The actual order for *E6* is 78, which implies a less dense packing than $E6'$, and this may have an analogy with the icosahedral group in which the icosahedron with $20 = 2 \times 10$ lattice sites has the same symmetry group as the dodecahedron with $12 = 2 \times 6$ lattice sites.

If *E8* is too small to represent a TOE with 10 or more dimensions, then we should consider the next larger "Exceptional" groups, such as *E10* or *E12*. Our prior *SU*(11) Boson GUT was a 10-dimensional representation, but our prior *SU*(13) and *SU*(15) Fermion GUT's were 12- and 14-dimensional representations. Furthermore, the *SU*(11) Boson GUT did

**Table 18 – A Pattern in Certain Exceptional Lie Algebras**

| $n$ | Name | $N$ = Order | Composition |
|---|---|---|---|
| 2 | $G2$ | 14 | $2 \times (2 \times 3 + 1)$ |
| 4 | $F4$ | 52 | $4 \times (2 \times 6 + 1)$ |
| 6 | $E6$ vs. $E6'$? | 78 vs. 126 | $6 \times (2 \times 6 + 1)$ vs. $6 \times (2 \times 10 + 1)$ |
| 8 | $E8$ | 248 | $8 \times (2 \times 15 + 1)$ |
| 10 | $E10$? | 430 | $10 \times (2 \times 21 + 1)$ |
| 12 | $E12$? | 684 | $12 \times (2 \times 28 + 1)$ |

not enumerate every degree of freedom. For instance, the Grand Bosons, *Q*, *R* and *U* all have color (c) and anti-color (a) degrees of freedom that were not individually stated. When we include all of these additional states, we find that our *SU*(11) GUT contains $204 = 12 \times 17$ boson degrees of freedom.

A full counting of the particle spectrum for Hyperflavor fermions with three generations of quarks and leptons, and two generations of leptoquarks, gives $504 = 18 \times 28 = 12 \times 42$ (see Section 7.4) fermion degrees of freedom. Note that there is an imbalance between the number of bosonic and fermionic degrees of freedom and, therefore, minimal Supersymmetry (SUSY) will produce sfermions that cannot be identified with existing bosons. Our 504-plet of fermions includes two 12-dimensional singlet states at the origin: $(\mathrm{a}_1, \psi_1)_{L,R;C,M,Y}$ and $(\mathrm{v}_1, \omega_1)_{L,R;C,M,Y}$. If we split this 504-plet into a 480-plet and two singlet 12-plets, then we can combine these $480 = 12 \times 40$ fermion degrees of freedom with $204 = 12 \times 17$ boson degrees of freedom into one $E12$ 684-plet ($684 = 12 \times (40 + 17)$).

Lisi introduced extra SUSY or "ghost" particles because he could not fill the $E8$ spectrum with known particles. Certainly, any non-minimal SUSY models (such as Hyper-SUSY in Section 7.6) can add significant confusion. This particle spectrum exactly fills one $E12$ plus two singlet 12-plets, however minimal SUSY requires an integer-spin sfermion $(\tilde{f})$ for every fermion $(f)$ and a half-integer-spin bosino $(\tilde{b})$ for every boson $(b)$. This requires two $E12$ representations, and the natural expectation is that one $E12$ contains bosons and sfermions, whereas the other $E12$ contains fermions and bosinos. Minimal SUSY may require singlet fermion groups that become basis vectors in the adjunct Supersymmetric representation. Table 19 is a deceptively simple list of bosons and fermions in one $E12$ representation.

## Table 19 – A Deceptively Simple *E12* TOE Particle Matrix

12-D "Basis" Bosons:

| | | | | | | | | | | | |
|---|---|---|---|---|---|---|---|---|---|---|---|
| $g_3$ | $g_8$ | $\gamma^0$ | $Z^0$ | $B_1^0$ | $B_2^0$ | $G$ | $F_3$ | $F_8$ | $F_{15}$ | $U_{10}^0$ | $U_{10}^{0*}$ |

$12 \times 56 = 672$ Roots of $E12$:

| | | | | | | | | | | | |
|---|---|---|---|---|---|---|---|---|---|---|---|
| $g_{GC}$ | $g_{BC}$ | $g_{RM}$ | $U_{7C}^{1/3}$ | $U_{8C}^{1/3}$ | $U_{9C}^{1/3}$ | $U_{7M}^{1/3}$ | $U_{8M}^{1/3}$ | $U_{9M}^{1/3}$ | $U_{7Y}^{1/3}$ | $U_{8Y}^{1/3}$ | $U_{9Y}^{1/3}$ |
| $g_{BM}$ | $g_{RY}$ | $g_{GY}$ | $U_{7R}^{-1/3}$ | $U_{8R}^{-1/3}$ | $U_{9R}^{-1/3}$ | $U_{7G}^{-1/3}$ | $U_{8G}^{-1/3}$ | $U_{9G}^{-1/3}$ | $U_{7B}^{-1/3}$ | $U_{8B}^{-1/3}$ | $U_{9B}^{-1/3}$ |
| $X_R^{5/3}$ | $X_G^{5/3}$ | $X_B^{5/3}$ | $Y_R^{-1/3}$ | $Y_G^{-1/3}$ | $Y_B^{-1/3}$ | $X_C^{-5/3}$ | $X_M^{-5/3}$ | $X_Y^{-5/3}$ | $Y_C^{1/3}$ | $Y_M^{1/3}$ | $Y_Y^{1/3}$ |
| $W_1^-$ | $W_1^+$ | $W_2^-$ | $W_2^+$ | $W_3^-$ | $W_3^+$ | $W_4^-$ | $W_4^+$ | $H_L$ | $\Phi_Z$ | $\Phi_{W_1}$ | $\Phi_{W_1}^*$ |
| $H_H$ | $H_P$ | $H^-$ | $H^+$ | $\Phi_{B_1}$ | $\Phi_{B_2}$ | $\Phi_{W_2}$ | $\Phi_{W_3}$ | $\Phi_{W_4}$ | $\Phi_{W_2}^*$ | $\Phi_{W_3}^*$ | $\Phi_{W_4}^*$ |
| $F_A$ | $F_B$ | $F_C$ | $F_D$ | $F_E$ | $F_F$ | $F_A^*$ | $F_R^*$ | $F_C^*$ | $F_D^*$ | $F_E^*$ | $F_F^*$ |
| $V_1^0$ | $V_2^0$ | $V_3^0$ | $V_4^0$ | $V_5^0$ | $V_6^0$ | $V_1^{0*}$ | $V_2^{0*}$ | $V_3^{0*}$ | $V_4^{0*}$ | $V_5^{0*}$ | $V_6^{0*}$ |
| $U_{1C}^{1/3}$ | $U_{2C}^{1/3}$ | $U_{3C}^{1/3}$ | $U_{4C}^{1/3}$ | $U_{5C}^{1/3}$ | $U_{6C}^{1/3}$ | $U_{1R}^{-1/3}$ | $U_{2R}^{-1/3}$ | $U_{3R}^{-1/3}$ | $U_{4R}^{-1/3}$ | $U_{5R}^{-1/3}$ | $U_{6R}^{-1/3}$ |
| $U_{1M}^{1/3}$ | $U_{2M}^{1/3}$ | $U_{3M}^{1/3}$ | $U_{4M}^{1/3}$ | $U_{5M}^{1/3}$ | $U_{6M}^{1/3}$ | $U_{1G}^{-1/3}$ | $U_{2G}^{-1/3}$ | $U_{3G}^{-1/3}$ | $U_{4G}^{-1/3}$ | $U_{5G}^{-1/3}$ | $U_{6G}^{-1/3}$ |
| $U_{1Y}^{1/3}$ | $U_{2Y}^{1/3}$ | $U_{3Y}^{1/3}$ | $U_{4Y}^{1/3}$ | $U_{5Y}^{1/3}$ | $U_{6Y}^{1/3}$ | $U_{1B}^{-1/3}$ | $U_{2B}^{-1/3}$ | $U_{3B}^{-1/3}$ | $U_{4B}^{-1/3}$ | $U_{5B}^{-1/3}$ | $U_{6B}^{-1/3}$ |
| $R_{1C}^{-1/6}$ | $R_{2C}^{-1/6}$ | $R_{3C}^{-1/6}$ | $R_{4C}^{-1/6}$ | $R_{5C}^{-1/6}$ | $R_{6C}^{-1/6}$ | $R_{1R}^{1/6}$ | $R_{2R}^{1/6}$ | $R_{3R}^{1/6}$ | $R_{4R}^{1/6}$ | $R_{5R}^{1/6}$ | $R_{6R}^{1/6}$ |
| $R_{1M}^{-1/6}$ | $R_{2M}^{-1/6}$ | $R_{3M}^{-1/6}$ | $R_{4M}^{-1/6}$ | $R_{5M}^{-1/6}$ | $R_{6M}^{-1/6}$ | $R_{1G}^{1/6}$ | $R_{2G}^{1/6}$ | $R_{3G}^{1/6}$ | $R_{4G}^{1/6}$ | $R_{5G}^{1/6}$ | $R_{6G}^{1/6}$ |
| $R_{1Y}^{-1/6}$ | $R_{2Y}^{-1/6}$ | $R_{3Y}^{-1/6}$ | $R_{4Y}^{-1/6}$ | $R_{5Y}^{-1/6}$ | $R_{6Y}^{-1/6}$ | $R_{1B}^{1/6}$ | $R_{2B}^{1/6}$ | $R_{3B}^{1/6}$ | $R_{4B}^{1/6}$ | $R_{5B}^{1/6}$ | $R_{6B}^{1/6}$ |
| $Q_{1R}^{1/6}$ | $Q_{2R}^{1/6}$ | $Q_{3R}^{1/6}$ | $Q_{4R}^{1/6}$ | $Q_{5R}^{1/6}$ | $Q_{6R}^{1/6}$ | $Q_{1C}^{-1/6}$ | $Q_{2C}^{-1/6}$ | $Q_{3C}^{-1/6}$ | $Q_{4C}^{-1/6}$ | $Q_{5C}^{-1/6}$ | $Q_{6C}^{-1/6}$ |
| $Q_{1G}^{1/6}$ | $Q_{2G}^{1/6}$ | $Q_{3G}^{1/6}$ | $Q_{4G}^{1/6}$ | $Q_{5G}^{1/6}$ | $Q_{6G}^{1/6}$ | $Q_{1M}^{-1/6}$ | $Q_{2M}^{-1/6}$ | $Q_{3M}^{-1/6}$ | $Q_{4M}^{-1/6}$ | $Q_{5M}^{-1/6}$ | $Q_{6M}^{-1/6}$ |
| $Q_{1B}^{1/6}$ | $Q_{2B}^{1/6}$ | $Q_{3B}^{1/6}$ | $Q_{4B}^{1/6}$ | $Q_{5B}^{1/6}$ | $Q_{6B}^{1/6}$ | $Q_{1Y}^{-1/6}$ | $Q_{2Y}^{-1/6}$ | $Q_{3Y}^{-1/6}$ | $Q_{4Y}^{-1/6}$ | $Q_{5Y}^{-1/6}$ | $Q_{6Y}^{-1/6}$ |
| $u_{1L,R}$ | $u_{1L,G}$ | $u_{1L,B}$ | $u_{1R,R}$ | $u_{1R,G}$ | $u_{1R,B}$ | $d_{1L,R}$ | $d_{1L,G}$ | $d_{1L,B}$ | $d_{1R,R}$ | $d_{1R,G}$ | $d_{1R,B}$ |
| $c_{1L,R}$ | $c_{1L,G}$ | $c_{1L,B}$ | $c_{1R,R}$ | $c_{1R,G}$ | $c_{1R,B}$ | $s_{1L,R}$ | $s_{1L,G}$ | $s_{1L,B}$ | $s_{1R,R}$ | $s_{1R,G}$ | $s_{1R,B}$ |
| $t_{1L,R}$ | $t_{1L,G}$ | $t_{1L,B}$ | $t_{1R,R}$ | $t_{1R,G}$ | $t_{1R,B}$ | $b_{1L,R}$ | $b_{1L,G}$ | $b_{1L,B}$ | $b_{1R,R}$ | $b_{1R,G}$ | $b_{1R,B}$ |
| $e_{1L}$ | $e_{1R}$ | $\nu_{e1L}$ | $\nu_{e1R}$ | $\mu_{1L}$ | $\mu_{1R}$ | $\nu_{\mu 1L}$ | $\nu_{\mu 1R}$ | $\tau_{1L}$ | $\tau_{1R}$ | $\nu_{\tau 1L}$ | $\nu_{\tau 1R}$ |
| $u_{2L,R}$ | $u_{2L,G}$ | $u_{2L,B}$ | $u_{2R,R}$ | $u_{2R,G}$ | $u_{2R,B}$ | $d_{2L,R}$ | $d_{2L,G}$ | $d_{2L,B}$ | $d_{2R,R}$ | $d_{2R,G}$ | $d_{2R,B}$ |
| $u_{3L,R}$ | $u_{3L,G}$ | $u_{3L,B}$ | $u_{3R,R}$ | $u_{3R,G}$ | $u_{3R,B}$ | $d_{3L,R}$ | $d_{3L,G}$ | $d_{3L,B}$ | $d_{3R,R}$ | $d_{3R,G}$ | $d_{3R,B}$ |
| $u_{4L,R}$ | $u_{4L,G}$ | $u_{4L,B}$ | $u_{4R,R}$ | $u_{4R,G}$ | $u_{4R,B}$ | $d_{4L,R}$ | $d_{4L,G}$ | $d_{4L,B}$ | $d_{4R,R}$ | $d_{4R,G}$ | $d_{4R,B}$ |
| $u_{5L,R}$ | $u_{5L,G}$ | $u_{5L,B}$ | $u_{5R,R}$ | $u_{5R,G}$ | $u_{5R,B}$ | $d_{5L,R}$ | $d_{5L,G}$ | $d_{5L,B}$ | $d_{5R,R}$ | $d_{5R,G}$ | $d_{5R,B}$ |
| $u_{6L,R}$ | $u_{6L,G}$ | $u_{6L,B}$ | $u_{6R,R}$ | $u_{6R,G}$ | $u_{6R,B}$ | $d_{6L,R}$ | $d_{6L,G}$ | $d_{6L,B}$ | $d_{6R,R}$ | $d_{6R,G}$ | $d_{6R,B}$ |
| $u_{7L,R}$ | $u_{7L,G}$ | $u_{7L,B}$ | $u_{7R,R}$ | $u_{7R,G}$ | $u_{7R,B}$ | $d_{7L,R}$ | $d_{7L,G}$ | $d_{7L,B}$ | $d_{7R,R}$ | $d_{7R,G}$ | $d_{7R,B}$ |
| $c_{2L,R}$ | $c_{2L,G}$ | $c_{2L,B}$ | $c_{2R,R}$ | $c_{2R,G}$ | $c_{2R,B}$ | $s_{2L,R}$ | $s_{2L,G}$ | $s_{2L,B}$ | $s_{2R,R}$ | $s_{2R,G}$ | $s_{2R,B}$ |
| $c_{3L,R}$ | $c_{3L,G}$ | $c_{3L,B}$ | $c_{3R,R}$ | $c_{3R,G}$ | $c_{3R,B}$ | $s_{3L,R}$ | $s_{3L,G}$ | $s_{3L,B}$ | $s_{3R,R}$ | $s_{3R,G}$ | $s_{3R,B}$ |
| $c_{4L,R}$ | $c_{4L,G}$ | $c_{4L,B}$ | $c_{4R,R}$ | $c_{4R,G}$ | $c_{4R,B}$ | $s_{4L,R}$ | $s_{4L,G}$ | $s_{4L,B}$ | $s_{4R,R}$ | $s_{4R,G}$ | $s_{4R,B}$ |
| $c_{5L,R}$ | $c_{5L,G}$ | $c_{5L,B}$ | $c_{5R,R}$ | $c_{5R,G}$ | $c_{5R,B}$ | $s_{5L,R}$ | $s_{5L,G}$ | $s_{5L,B}$ | $s_{5R,R}$ | $s_{5R,G}$ | $s_{5R,B}$ |

| | | | | | | | | | | | |
|---|---|---|---|---|---|---|---|---|---|---|---|
| $c_{6L,R}$ | $c_{6L,G}$ | $c_{6L,B}$ | $c_{6R,R}$ | $c_{6R,G}$ | $c_{6R,B}$ | $s_{6L,R}$ | $s_{6L,G}$ | $s_{6L,B}$ | $s_{6R,R}$ | $s_{6R,G}$ | $s_{6R,B}$ |
| $c_{7L,R}$ | $c_{7L,G}$ | $c_{7L,B}$ | $c_{7R,R}$ | $c_{7R,G}$ | $c_{7R,B}$ | $s_{7L,R}$ | $s_{7L,G}$ | $s_{7L,B}$ | $s_{7R,R}$ | $s_{7R,G}$ | $s_{7R,B}$ |
| $t_{2L,R}$ | $t_{2L,G}$ | $t_{2L,B}$ | $t_{2R,R}$ | $t_{2R,G}$ | $t_{2R,B}$ | $b_{2L,R}$ | $b_{2L,G}$ | $b_{2L,B}$ | $b_{2R,R}$ | $b_{2R,G}$ | $b_{2R,B}$ |
| $t_{3L,R}$ | $t_{3L,G}$ | $t_{3L,B}$ | $t_{3R,R}$ | $t_{3R,G}$ | $t_{3R,B}$ | $b_{3L,R}$ | $b_{3L,G}$ | $b_{3L,B}$ | $b_{3R,R}$ | $b_{3R,G}$ | $b_{3R,B}$ |
| $t_{4L,R}$ | $t_{4L,G}$ | $t_{4L,B}$ | $t_{4R,R}$ | $t_{4R,G}$ | $t_{4R,B}$ | $b_{4L,R}$ | $b_{4L,G}$ | $b_{4L,B}$ | $b_{4R,R}$ | $b_{4R,G}$ | $b_{4R,B}$ |
| $t_{5L,R}$ | $t_{5L,G}$ | $t_{5L,B}$ | $t_{5R,R}$ | $t_{5R,G}$ | $t_{5R,B}$ | $b_{5L,R}$ | $b_{5L,G}$ | $b_{5L,B}$ | $b_{5R,R}$ | $b_{5R,G}$ | $b_{5R,B}$ |
| $t_{6L,R}$ | $t_{6L,G}$ | $t_{6L,B}$ | $t_{6R,R}$ | $t_{6R,G}$ | $t_{6R,B}$ | $b_{6L,R}$ | $b_{6L,G}$ | $b_{6L,B}$ | $b_{6R,R}$ | $b_{6R,G}$ | $b_{6R,B}$ |
| $t_{7L,R}$ | $t_{7L,G}$ | $t_{7L,B}$ | $t_{7R,R}$ | $t_{7R,G}$ | $t_{7R,B}$ | $b_{7L,R}$ | $b_{7L,G}$ | $b_{7L,B}$ | $b_{7R,R}$ | $b_{7R,G}$ | $b_{7R,B}$ |
| $e_{2L}$ | $e_{2R}$ | $\nu_{e2L}$ | $\nu_{e2R}$ | $\mu_{2L}$ | $\mu_{2R}$ | $\nu_{\mu 2L}$ | $\nu_{\mu 2R}$ | $\tau_{2L}$ | $\tau_{2R}$ | $\nu_{\tau 2L}$ | $\nu_{\tau 2R}$ |
| $e_{3L}$ | $e_{3R}$ | $\nu_{e3L}$ | $\nu_{e3R}$ | $\mu_{3L}$ | $\mu_{3R}$ | $\nu_{\mu 3L}$ | $\nu_{\mu 3R}$ | $\tau_{3L}$ | $\tau_{3R}$ | $\nu_{\tau 3L}$ | $\nu_{\tau 3R}$ |
| $e_{4L}$ | $e_{4R}$ | $\nu_{e4L}$ | $\nu_{e4R}$ | $\mu_{4L}$ | $\mu_{4R}$ | $\nu_{\mu 4L}$ | $\nu_{\mu 4R}$ | $\tau_{4L}$ | $\tau_{4R}$ | $\nu_{\tau 4L}$ | $\nu_{\tau 4R}$ |
| $e_{5L}$ | $e_{5R}$ | $\nu_{e5L}$ | $\nu_{e5R}$ | $\mu_{5L}$ | $\mu_{5R}$ | $\nu_{\mu 5L}$ | $\nu_{\mu 5R}$ | $\tau_{5L}$ | $\tau_{5R}$ | $\nu_{\tau 5L}$ | $\nu_{\tau 5R}$ |
| $e_{6L}$ | $e_{6R}$ | $\nu_{e6L}$ | $\nu_{e6R}$ | $\mu_{6L}$ | $\mu_{6R}$ | $\nu_{\mu 6L}$ | $\nu_{\mu 6R}$ | $\tau_{6L}$ | $\tau_{6R}$ | $\nu_{\tau 6L}$ | $\nu_{\tau 6R}$ |
| $e_{7L}$ | $e_{7R}$ | $\nu_{e7L}$ | $\nu_{e7R}$ | $\mu_{7L}$ | $\mu_{7R}$ | $\nu_{\mu 7L}$ | $\nu_{\mu 7R}$ | $\tau_{7L}$ | $\tau_{7R}$ | $\nu_{\tau 7L}$ | $\nu_{\tau 7R}$ |
| $\mathrm{a}_{2L,C}$ | $\mathrm{a}_{2L,M}$ | $\mathrm{a}_{2L,Y}$ | $\mathrm{a}_{2R,C}$ | $\mathrm{a}_{2R,M}$ | $\mathrm{a}_{2R,Y}$ | $\psi_{2L,C}$ | $\psi_{2L,M}$ | $\psi_{2L,Y}$ | $\psi_{2R,C}$ | $\psi_{2R,M}$ | $\psi_{2R,Y}$ |
| $\mathrm{a}_{3L,C}$ | $\mathrm{a}_{3L,M}$ | $\mathrm{a}_{3L,Y}$ | $\mathrm{a}_{3R,C}$ | $\mathrm{a}_{3R,M}$ | $\mathrm{a}_{3R,Y}$ | $\psi_{3L,C}$ | $\psi_{3L,M}$ | $\psi_{3L,Y}$ | $\psi_{3R,C}$ | $\psi_{3R,M}$ | $\psi_{3R,Y}$ |
| $\mathrm{a}_{4L,C}$ | $\mathrm{a}_{4L,M}$ | $\mathrm{a}_{4L,Y}$ | $\mathrm{a}_{4R,C}$ | $\mathrm{a}_{4R,M}$ | $\mathrm{a}_{4R,Y}$ | $\psi_{4L,C}$ | $\psi_{4L,M}$ | $\psi_{4L,Y}$ | $\psi_{4R,C}$ | $\psi_{4R,M}$ | $\psi_{4R,Y}$ |
| $\mathrm{a}_{5L,C}$ | $\mathrm{a}_{5L,M}$ | $\mathrm{a}_{5L,Y}$ | $\mathrm{a}_{5R,C}$ | $\mathrm{a}_{5R,M}$ | $\mathrm{a}_{5R,Y}$ | $\psi_{5L,C}$ | $\psi_{5L,M}$ | $\psi_{5L,Y}$ | $\psi_{5R,C}$ | $\psi_{5R,M}$ | $\psi_{5R,Y}$ |
| $\mathrm{a}_{6L,C}$ | $\mathrm{a}_{6L,M}$ | $\mathrm{a}_{6L,Y}$ | $\mathrm{a}_{6R,C}$ | $\mathrm{a}_{6R,M}$ | $\mathrm{a}_{6R,Y}$ | $\psi_{6L,C}$ | $\psi_{6L,M}$ | $\psi_{6L,Y}$ | $\psi_{6R,C}$ | $\psi_{6R,M}$ | $\psi_{6R,Y}$ |
| $\mathrm{a}_{7L,C}$ | $\mathrm{a}_{7L,M}$ | $\mathrm{a}_{7L,Y}$ | $\mathrm{a}_{7R,C}$ | $\mathrm{a}_{7R,M}$ | $\mathrm{a}_{7R,Y}$ | $\psi_{7L,C}$ | $\psi_{7L,M}$ | $\psi_{7L,Y}$ | $\psi_{7R,C}$ | $\psi_{7R,M}$ | $\psi_{7R,Y}$ |
| $\mathrm{v}_{2L,C}$ | $\mathrm{v}_{2L,M}$ | $\mathrm{v}_{2L,Y}$ | $\mathrm{v}_{2R,C}$ | $\mathrm{v}_{2R,M}$ | $\mathrm{v}_{2R,Y}$ | $\omega_{2L,C}$ | $\omega_{2L,M}$ | $\omega_{2L,Y}$ | $\omega_{2R,C}$ | $\omega_{2R,M}$ | $\omega_{2R,Y}$ |
| $\mathrm{v}_{3L,C}$ | $\mathrm{v}_{3L,M}$ | $\mathrm{v}_{3L,Y}$ | $\mathrm{v}_{3R,C}$ | $\mathrm{v}_{3R,M}$ | $\mathrm{v}_{3R,Y}$ | $\omega_{3L,C}$ | $\omega_{3L,M}$ | $\omega_{3L,Y}$ | $\omega_{3R,C}$ | $\omega_{3R,M}$ | $\omega_{3R,Y}$ |
| $\mathrm{v}_{4L,C}$ | $\mathrm{v}_{4L,M}$ | $\mathrm{v}_{4L,Y}$ | $\mathrm{v}_{4R,C}$ | $\mathrm{v}_{4R,M}$ | $\mathrm{v}_{4R,Y}$ | $\omega_{4L,C}$ | $\omega_{4L,M}$ | $\omega_{4L,Y}$ | $\omega_{4R,C}$ | $\omega_{4R,M}$ | $\omega_{4R,Y}$ |
| $\mathrm{v}_{5L,C}$ | $\mathrm{v}_{5L,M}$ | $\mathrm{v}_{5L,Y}$ | $\mathrm{v}_{5R,C}$ | $\mathrm{v}_{5R,M}$ | $\mathrm{v}_{5R,Y}$ | $\omega_{5L,C}$ | $\omega_{5L,M}$ | $\omega_{5L,Y}$ | $\omega_{5R,C}$ | $\omega_{5R,M}$ | $\omega_{5R,Y}$ |
| $\mathrm{v}_{6L,C}$ | $\mathrm{v}_{6L,M}$ | $\mathrm{v}_{6L,Y}$ | $\mathrm{v}_{6R,C}$ | $\mathrm{v}_{6R,M}$ | $\mathrm{v}_{6R,Y}$ | $\omega_{6L,C}$ | $\omega_{6L,M}$ | $\omega_{6L,Y}$ | $\omega_{6R,C}$ | $\omega_{6R,M}$ | $\omega_{6R,Y}$ |
| $\mathrm{v}_{7L,C}$ | $\mathrm{v}_{7L,M}$ | $\mathrm{v}_{7L,Y}$ | $\mathrm{v}_{7R,C}$ | $\mathrm{v}_{7R,M}$ | $\mathrm{v}_{7R,Y}$ | $\omega_{7L,C}$ | $\omega_{7L,M}$ | $\omega_{7L,Y}$ | $\omega_{7R,C}$ | $\omega_{7R,M}$ | $\omega_{7R,Y}$ |

Two 12-D Singlets:

| | | | | | | | | | | | |
|---|---|---|---|---|---|---|---|---|---|---|---|
| $\mathrm{a}_{1L,C}$ | $\mathrm{a}_{1L,M}$ | $\mathrm{a}_{1L,Y}$ | $\mathrm{a}_{1R,C}$ | $\mathrm{a}_{1R,M}$ | $\mathrm{a}_{1R,Y}$ | $\psi_{1L,C}$ | $\psi_{1L,M}$ | $\psi_{1L,Y}$ | $\psi_{1R,C}$ | $\psi_{1R,M}$ | $\psi_{1R,Y}$ |
| $\mathrm{v}_{1L,C}$ | $\mathrm{v}_{1L,M}$ | $\mathrm{v}_{1L,Y}$ | $\mathrm{v}_{1R,C}$ | $\mathrm{v}_{1R,M}$ | $\mathrm{v}_{1R,Y}$ | $\omega_{1L,C}$ | $\omega_{1L,M}$ | $\omega_{1L,Y}$ | $\omega_{1R,C}$ | $\omega_{1R,M}$ | $\omega_{1R,Y}$ |

Table 19 is "deceptively simple" because the true $E12$ groups should contain bosons ($g,\gamma,Z,W,etc.$) and sfermions ($\tilde{u},\tilde{d},\tilde{e},\tilde{\nu}_e,\tilde{c},\tilde{t},etc.$) or fermions ($u,d,e,\nu_e,c,t,etc.$) and bosinos ($\tilde{g},\tilde{\gamma},\tilde{Z},\tilde{W},etc.$).

Note that $E12$ has $672 = 12 \times 56 = 24 \times 28$ roots and, thus, it naturally contains Hyperflavor $SO(8)$ 28-plets. Ten of these "Basis" Bosons lie on the diagonal of the $SU(11)$ Boson GUT and are directly related to the charges: $g_3$, $g_8$, $Y_L$ (or $Q$), $T_{3L}$, $T_{3HF}$, $T_{8HF}$, $m$, $F_3$, $F_8$ and $F_{15}$. The extra two basis vectors $\mathrm{Re}\,U_{10}^0$ and $\mathrm{Im}\,U_{10}^0$ are based on the $U_{10}^0$ and $U_{10}^{0*}$ bosons, which are related to the $\left(V_{1-3}^0, V_{1-3}^{0*}\right)$ and $\left(V_{4-6}^0, V_{4-6}^{0*}\right)$ bosons – these bosons all act like step-up/ step-down operators that transform one generation of leptoquarks, quarks, or neutrinos, respectively, into another. These bosonic operators have a similar effect as Lisi's triality, and are important in the CKM-PMNS-Munroe Mixing Matrix.

How should this $E12$ decompose? The expectation is that these twelve dimensions break into a 6-dimensional $E6'_{Color/HEW}$ (Spacetime and the Flavorbrane/ Weakbrane) and another 6-dimensional $E6'_{Gravity/CKM}$ (Gravity, the Gravitybrane and the first heirarchal 2-brane). The results are listed in Table 20. Note that some particle states are not individually listed because $E6' \times E6'$ is smaller than $E12$. In Table 20, the $U_{10}^0$ boson is closely associated with the graviton and WIMP-gravitons, which are spin-2 tensor bosons. If $U_{10}^0$ is also a spin-2 tensor boson, then this introduces the possibility of spin-3/2 gravitino-like $\psi$ and $\omega$ leptoquarks, which would require another representation group for these particles.

How does $E6'_{Color/HEW}$ decompose? The underlying string symmetries imply $6 = 4 + 2$ dimensions, but the most interesting subgroups of $E6'$ imply an inconsistent composition of $6 = 2 + 4$. This composition is interesting as a pedagogical tool but should, at best, represent a symbolic TOE. This decomposition of $E6'_{Color/HEW} \to G2_{Color} \times F4_{HEW}$ is symbolically reviewed in Table 21. Note that the $G2_{Color}$ sub-TOE is based on a

**Table 20 – The Decomposition of *E12*:**

$E6'$ Color/ HEW

6-D "Basis" Bosons:

| | | | | | |
|---|---|---|---|---|---|
| $g_3$ | $g_8$ | $\gamma^0$ | $Z^0$ | $B_1^0$ | $B_2^0$ |

$6 \times 20 = 120$ Roots:

| | | | | | |
|---|---|---|---|---|---|
| $g_{RM}$ | $g_{RY}$ | $g_{GC}$ | $g_{GY}$ | $g_{BC}$ | $g_{BM}$ |
| $X_R^{5/3}$ | $X_G^{5/3}$ | $X_B^{5/3}$ | $Y_R^{-1/3}$ | $Y_G^{-1/3}$ | $Y_B^{-1/3}$ |
| $X_C^{-5/3}$ | $X_M^{-5/3}$ | $X_Y^{-5/3}$ | $Y_C^{1/3}$ | $Y_M^{1/3}$ | $Y_Y^{1/3}$ |
| $W_1^-$ | $W_1^+$ | $H_L$ | $H_H$ | $\Phi_{B_1}$ | $\Phi_{B_2}$ |
| $W_2^-$ | $W_2^+$ | $\Phi_Z$ | $H_P$ | $\Phi_{W_2}$ | $\Phi_{W_2}^*$ |
| $W_3^-$ | $W_3^+$ | $\Phi_{W_1}$ | $H^-$ | $\Phi_{W_3}$ | $\Phi_{W_3}^*$ |
| $W_4^-$ | $W_4^+$ | $\Phi_{W_1}^*$ | $H^+$ | $\Phi_{W_4}$ | $\Phi_{W_4}^*$ |
| $d_{1L,R}$ | $d_{1L,G}$ | $d_{1L,B}$ | $d_{1R,R}$ | $d_{1R,G}$ | $d_{1R,B}$ |
| $u_{2L,R}$ | $u_{2L,G}$ | $u_{2L,B}$ | $u_{2R,R}$ | $u_{2R,G}$ | $u_{2R,B}$ |
| $d_{2L,R}$ | $d_{2L,G}$ | $d_{2L,B}$ | $d_{2R,R}$ | $d_{2R,G}$ | $d_{2R,B}$ |
| $u_{3L,R}$ | $u_{3L,G}$ | $u_{3L,B}$ | $u_{3R,R}$ | $u_{3R,G}$ | $u_{3R,B}$ |
| $d_{3L,R}$ | $d_{3L,G}$ | $d_{3L,B}$ | $d_{3R,R}$ | $d_{3R,G}$ | $d_{3R,B}$ |
| $u_{4L,R}$ | $u_{4L,G}$ | $u_{4L,B}$ | $u_{4R,R}$ | $u_{4R,G}$ | $u_{4R,B}$ |
| $d_{4L,R}$ | $d_{4L,G}$ | $d_{4L,B}$ | $d_{4R,R}$ | $d_{4R,G}$ | $d_{4R,B}$ |
| $u_{5L,R}$ | $u_{5L,G}$ | $u_{5L,B}$ | $u_{5R,R}$ | $u_{5R,G}$ | $u_{5R,B}$ |
| $d_{5L,R}$ | $d_{5L,G}$ | $d_{5L,B}$ | $d_{5R,R}$ | $d_{5R,G}$ | $d_{5R,B}$ |
| $u_{6L,R}$ | $u_{6L,G}$ | $u_{6L,B}$ | $u_{6R,R}$ | $u_{6R,G}$ | $u_{6R,B}$ |
| $d_{6L,R}$ | $d_{6L,G}$ | $d_{6L,B}$ | $d_{6R,R}$ | $d_{6R,G}$ | $d_{6R,B}$ |
| $u_{7L,R}$ | $u_{7L,G}$ | $u_{7L,B}$ | $u_{7R,R}$ | $u_{7R,G}$ | $u_{7R,B}$ |
| $d_{7L,R}$ | $d_{7L,G}$ | $d_{7L,B}$ | $d_{7R,R}$ | $d_{7R,G}$ | $d_{7R,B}$ |

6-D Singlet:

| | | | | | |
|---|---|---|---|---|---|
| $u_{1L,R}$ | $u_{1L,G}$ | $u_{1L,B}$ | $u_{1R,R}$ | $u_{1R,G}$ | $u_{1R,B}$ |

$E6'$ Gravity/ CKM

6-D "Basis" Bosons:

| | | | | | |
|---|---|---|---|---|---|
| $G$ | $F_3$ | $F_8$ | $F_{15}$ | $U_{10}^0$ | $U_{10}^{0*}$ |

$6 \times 20 = 120$ Roots:

| | | | | | |
|---|---|---|---|---|---|
| $F_A$ | $F_B$ | $F_C$ | $F_D$ | $F_E$ | $F_F$ |
| $F_A^*$ | $F_B^*$ | $F_C^*$ | $F_D^*$ | $F_E^*$ | $F_F^*$ |
| $V_1^0$ | $V_2^0$ | $V_3^0$ | $V_4^0$ | $V_5^0$ | $V_6^0$ |
| $V_1^{0*}$ | $V_2^{0*}$ | $V_3^{0*}$ | $V_4^{0*}$ | $V_5^{0*}$ | $V_6^{0*}$ |
| $U_{1c}^{-1/3}$ | $U_{2c}^{-1/3}$ | $U_{3c}^{-1/3}$ | $U_{4c}^{-1/3}$ | $U_{5c}^{-1/3}$ | $U_{6c}^{-1/3}$ |
| $U_{1a}^{1/3}$ | $U_{2a}^{1/3}$ | $U_{3a}^{1/3}$ | $U_{4a}^{1/3}$ | $U_{5a}^{1/3}$ | $U_{6a}^{1/3}$ |
| $U_{7c}^{-1/3}$ | $U_{8c}^{-1/3}$ | $U_{9c}^{-1/3}$ | $U_{7a}^{1/3}$ | $U_{8a}^{1/3}$ | $U_{9a}^{1/3}$ |
| $R_{1c}^{1/6}$ | $R_{2c}^{1/6}$ | $R_{3c}^{1/6}$ | $R_{4c}^{1/6}$ | $R_{5c}^{1/6}$ | $R_{6c}^{1/6}$ |
| $R_{1a}^{-1/6}$ | $R_{2a}^{-1/6}$ | $R_{3a}^{-1/6}$ | $R_{4a}^{-1/6}$ | $R_{5a}^{-1/6}$ | $R_{6a}^{-1/6}$ |
| $Q_{1c}^{1/6}$ | $Q_{2c}^{1/6}$ | $Q_{3c}^{1/6}$ | $Q_{4c}^{1/6}$ | $Q_{5c}^{1/6}$ | $Q_{6c}^{1/6}$ |
| $Q_{1a}^{-1/6}$ | $Q_{2a}^{-1/6}$ | $Q_{3a}^{-1/6}$ | $Q_{4a}^{-1/6}$ | $Q_{5a}^{-1/6}$ | $Q_{6a}^{-1/6}$ |
| $s_{1L,R}$ | $s_{1L,G}$ | $s_{1L,B}$ | $s_{1R,R}$ | $s_{1R,G}$ | $s_{1R,B}$ |
| $b_{1L,R}$ | $b_{1L,G}$ | $b_{1L,B}$ | $b_{1R,R}$ | $b_{1R,G}$ | $b_{1R,B}$ |
| $\nu_{e1L}$ | $\nu_{e1R}$ | $\nu_{\mu 1L}$ | $\nu_{\mu 1R}$ | $\nu_{\tau 1L}$ | $\nu_{\tau 1R}$ |
| $\psi_{1L,C}$ | $\psi_{1L,M}$ | $\psi_{1L,Y}$ | $\psi_{1R,C}$ | $\psi_{1R,M}$ | $\psi_{1R,Y}$ |
| $\omega_{1L,C}$ | $\omega_{1L,M}$ | $\omega_{1L,Y}$ | $\omega_{1R,C}$ | $\omega_{1R,M}$ | $\omega_{1R,Y}$ |
| $u_{1L,R}$ | $u_{1L,G}$ | $u_{1L,B}$ | $u_{1R,R}$ | $u_{1R,G}$ | $u_{1R,B}$ |
| $c_{1L,R}$ | $c_{1L,G}$ | $c_{1L,B}$ | $c_{1R,R}$ | $c_{1R,G}$ | $c_{1R,B}$ |
| $t_{1L,R}$ | $t_{1L,G}$ | $t_{1L,B}$ | $t_{1R,R}$ | $t_{1R,G}$ | $t_{1R,B}$ |
| $e_{1L}$ | $e_{1R}$ | $\mu_{1L}$ | $\mu_{1R}$ | $\tau_{1L}$ | $\tau_{1R}$ |

6-D Singlet:

| | | | | | |
|---|---|---|---|---|---|
| $d_{1L,R}$ | $d_{1L,G}$ | $d_{1L,B}$ | $d_{1R,R}$ | $d_{1R,G}$ | $d_{1R,B}$ |

2-dimensional trianglar close-packing lattice with 6 nearest neighbors that defines color, quarks and leptons. And the $F4_{HEW}$ sub-TOE is based on a 3-dimensional tetrahedral close-packing lattice with 12 nearest neighbors that defines Hyperflavor-Electroweak with weak isospin and Hyperflavor 7-plets.

**Table 21 – The Decomposition of an *E6′* of Color/ HEW:**

G2 Color: F4 HEW:

| Basis: | |
|---|---|
| $g_3$ | $g_8$ |
| **Roots:** | |
| $g_{RM}$ | $u_{1L,R}$ |
| $g_{RY}$ | $u_{1L,G}$ |
| $g_{GC}$ | $u_{1L,B}$ |
| $g_{GY}$ | $u_{1R,R}$ |
| $g_{BC}$ | $u_{1R,G}$ |
| $g_{BM}$ | $u_{1R,B}$ |
| **Singlet:** | |
| $e_{1L}$ | $e_{1R}$ |

| "Basis" Bosons: | | | |
|---|---|---|---|
| $\gamma^0$ | $Z^0$ | $B_1^0$ | $B_2^0$ |
| $4\times 12=48$ Roots: | | | |
| $W_1^-$ | $W_2^-$ | $W_3^-$ | $W_4^-$ |
| $W_1^+$ | $W_2^+$ | $W_3^+$ | $W_4^+$ |
| $H_L$ | $\Phi_Z$ | $\Phi_{W_1}$ | $\Phi_{W_1}^*$ |
| $H_H$ | $H_P$ | $H^-$ | $H^+$ |
| $\Phi_{B_1}$ | $\Phi_{W_2}$ | $\Phi_{W_3}$ | $\Phi_{W_4}$ |
| $\Phi_{B_2}$ | $\Phi_{W_2}^*$ | $\Phi_{W_3}^*$ | $\Phi_{W_4}^*$ |
| $e_{2L}$ | $e_{2R}$ | $\nu_{e2L}$ | $\nu_{e2R}$ |
| $e_{3L}$ | $e_{3R}$ | $\nu_{e3L}$ | $\nu_{e3R}$ |
| $e_{4L}$ | $e_{4R}$ | $\nu_{e4L}$ | $\nu_{e4R}$ |
| $e_{5L}$ | $e_{5R}$ | $\nu_{e5L}$ | $\nu_{e5R}$ |
| $e_{6L}$ | $e_{6R}$ | $\nu_{e6L}$ | $\nu_{e6R}$ |
| $e_{7L}$ | $e_{7R}$ | $\nu_{e7L}$ | $\nu_{e7R}$ |
| 4-D Singlet: | | | |
| $e_{1L}$ | $e_{1R}$ | $\nu_{e1L}$ | $\nu_{e1R}$ |

If we assume the necessity of singlet fermion states in a Supersymmetric Exceptional Lie Algebra representation, then we realize two important conclusions: 1) Leptons are a color singlet – the "color" White (see Table 21 and Section 7.4), and 2) the $F4_{HEW}$ consists of a $D4=SO(7,1)$ 28-plet of bosons, and a matching $D4=SO(7,1)$ 28-plet of fermions. This $D4$ of fermions has the seven-fold $(7\times 4)$ "septality" symmetry of Hyperflavor, and can be interpreted as Lisi's three-fold $(3\times 8)$ triality symmetry with the omission of the fermion singlet state.

The next natural extension of $E12$ is $E14$, which should have an order of $14\times(2\times 36+1)=1022$, including $14\times 72=28\times 36=1008$ roots – exactly 50% more roots than $E12$. However, the bosons of $SU(11)$ do not fit naturally into the 14-dimensional symmetry of $E14$.

## 7.8 The Crossroads of Science and Religion?

There is some interesting numerology embedded in this book's analysis of the Universe. For instance, Dirac's Large Number is $\sim 10^{40}$, and forty is the Judeo-Christian number for "perfect testing", i.e. Noah and the 40 days of flood (Genesis 7:12), Moses and the Israelites in the desert for 40 years (Exodus 16:35), and Jesus in the desert for 40 days (Mark 1:13). Various powers of this number and its inverse appear throughout this book. This book also estimated the total number of degrees of freedom on a 10-dimensional string as $\sim \left(10^{40}\right)^{12}$ and used an $E12$ TOE, and twelve is another special grouping number in Judeo-Christian theology, i.e. the 12 tribes of Israel (Genesis 49:28), the 12 apostles of Jesus (Mark 3:14), and the 12 gates of New Jerusalem (Revelations 21:12). Similarly, this theory has seven Hyperflavor types of leptons, quarks and leptoquarks $\left(e_{1-7}, \nu_{e1-7}, u_{1-7}, d_{1-7}, \mathrm{a}_{1-7}, \psi_{1-7}, \text{etc.}\right)$; and seven is the Judeo-Christian number of "perfect completion", i.e. God rested on the seventh day (Genesis 2:2-3), required the Sabbath on the seventh day (Exodus 16:26-30), and the Book of Revelations contains many sevens and seven-halves. This book also has three strong gauge forces, three low mass generations of fermions, and three triplets of space/ string membranes on a ten-dimensional string; and three is the number of the Trinity of God – the Father, Son and Holy Spirit. Also, the stella octangula (two nested 3-simplices) in Section 7.2 is the three-dimensional extension of the two-dimensional Star of David (two nested 2-simplices).

Other cultures hold high regard for some of these numbers. For instance, the Cherokee Native Americans consider seven and four to be special numbers. There are seven Cherokee clans: long hair, blue, wolf, wild pototo, deer, bird and paint. And there are four cardinal directions:

east, west, north and south. This theory also has tetrahedra of leptons (or quarks) and matter; each of which has 4 vertices; and each of which reduces to 4 important directions after symmetry-breaking: $(\pm \mathbf{T}_{3L}, \pm \mathbf{T}_{3R})$ and $(\pm \mathbf{T}_{3WG}, \pm \mathbf{T}_{3i})$, respectively. The product of these numbers, 28, is also important to this theory: 28 HEW bosons, 28 Hyperflavor leptons (or quarks, not counting color) per generation, and 2 × 28 Grand Bosons (*Q, R, U* and *V*). There are also similarities to the calendar: seven days in a week, about four weeks (~ 4 × 7 ~ 28 days) in a lunar cycle (actual length = 29.53 days), and the Babylonians chose 360° in a circle (similar to *SU*(19)) – because there are 365.24 days per year.

The Anthropic Principle states that the fundamental parameters of the Universe have been fine-tuned to allow life to develop and prosper [48]. The paradox is that it requires intelligent life to pose this principle regarding the necessary conditions for life – hundreds of millions of years after the fact. Some modern science writers have implied that these parameters are completely random – that, for instance, a Universe might exist wherein there are only two spatial dimensions and any life-forms must excrete digested food waste out of their mouths; or a Universe might exist wherein the strong force is weak enough that Hydrogen is the only stable element, and no complex organic bonds can form.

In the *Creation* versus *Design* versus *Chance* versus *Necessity* debate, the Anthropic Principle then leads to either 1) *Creation by God* such as the creation(s) described in the beginning of Genesis (We were specifically created for a relationship with God – God created the heavens and earth, spoke "Let there be light", created everything in six days, created Adam out of dust, and created Eve from Adam's rib – this is often interpreted literally despite the fact that Jesus taught with metaphors), 2) an *Intelligent Designer* who set these parameters (a Supreme Being *coveted life throughout the Universe* – This assumption is free of the literalism and dogma that plagues the first, but does not necessarily imply the same God (examples: this Designer could be consistent with the God of the Bible – a God who desires to have a relationship with us, *OR* Aristotle's Prime Mover – an uninvolved Creator), although the Bible does contain design-friendly language: "For since the creation of the world God's invisible qualities – His eternal power and divine nature – have been clearly seen, being understood from what has been made, so that men are without excuse" (Romans 1:20)); or 3) a *Random Multiverse* consisting of a large, *mostly unobservable*, Random Chance sampling of these possible Universes (*a cosmic Las Vegas* with a few winning, and many more losing, Universes – Life is an accidental organization of stardust that, like all stars, will burn out with little or no impact); or 4) a GUT that *necessitates* these parameters (the Universe *obeys physical laws that sanction life* – just as "one plus one equals two" (in our familiar base ten), the inevitability of complex life may be a fundamental property of Nature).

These paradigms each have quirky spin-off conclusions. If an Intelligent Designer formed the entire Universe to have parameters desirable for life, why would the Designer stop with just Earth? Does the Intelligent Designer expect mankind's descendants to someday fill the Universe or does alien life (whom the Designer also loves) already fill the Universe? When we consider that there may be an unknowable number of Universes, each of which has a potential for billions of Galaxies, each of which has a potential for billions of stars and an untold number of inhabitable planets; are we limiting an Almighty God too much to expect only one living planet? Is it more reasonable to expect a loving, Almighty God to have made life on one planet via Creation, or on many planets via a life-preferring GUT and hundreds of millions of years of evolution? This question may speak more to God's intent and purpose for life. Besides, "Let there be light" may be a perfect metaphor for the Big Bang (as well as the later recombination event that birthed the cosmic microwave background radiation). And the Chance paradigm requires nearly as much *Faith* as does Creation or Design – Do we believe in unseen Universes or an unseen God or Designer? Certainly, an unseen Designer implies a higher level of unseen complexity than do unseen Universes, and thus may require more faith, but both options require believing in something that we cannot yet prove scientifically. The Necessity paradigm may yet be a sub-paradigm of either of the others, i.e. 1) God (or the Designer) chose a GUT and manipulated its laws to create life, or 2) the GUT allows only a subset of all imaginable "Random Chance" Universes to exist and a few of these failed Universes may have never inflated much past the Planck size, or may have collapsed into Black Holes.

However, this theory suggests that underlying mathematical symmetries caused the potentially complex String landscape to break into a spacetime Universe and a possible hierarchy of string/ m-brane dimensions, and that Thermodynamic and Energy-Mass equilibrium considerations (Equations (8), (18) or (35), and (32)) predisposed the coupling strengths to attain their modern-day values. Thus, there may not be very many permutations of the Universe that are mathematically consistent. Can Science ever prove the existence of an Intelligent Designer? If so, can Science prove the Designer's characteristics? Or will a "leap of faith" always be required by religions? Decide for yourself if this Grand Unified Theory has implications in the Creation versus Design versus Chance versus Necessity debate.

*God (Nature) used beautiful mathematics in creating the world.* — Paul Dirac

*God does not play dice.* — Albert Einstein

# 8. Summary

This book contains several major paradigm-changing concepts. The most original is the Quantum Statistical Grand Unification (QSGUT) of the low-energy couplings of the four fundamental forces. Except for the Strong Nuclear force, this pattern can be stated as:

$$\alpha_n = C_1 M_n n^{(D-4)} \Big/ \left\{ \exp\left[\beta^4 \left(n^4 \mathcal{Q}_1 - \mu^4\right)\right] - 1 \right\} \tag{8}$$

where $\alpha_2$ is the Electromagnetic coupling between two electrons and/ or protons, $\alpha_3$ is the nucleonic Weak Nuclear coupling at a momentum-transfer-scale equal to the electron mass, and $\alpha_4$ is the gravitational coupling between two protons. The original approximation was one of independent, non-interacting GUM Bosons (Section 4.1), but non-linear (renormalization) corrections were later considered in the low-energy coupling of the Strong Nuclear force (Section 5.3).

The second important idea examines the overlapping physics between QSGUT and the Renormalization Group Equations, and leads to Variable Coupling Theory (VCT) (Sections 5.1 and 6.2):

$$d(\ln \alpha_{n\leq 2}) / d(\ln \mu) = -\, b_n \alpha_n / 2\pi, \text{ and} \tag{25}$$

$$d(\ln \alpha_{n\geq 3}) / d(\ln \mu) = 2\left(n^4 - 16\right) / 65 \tag{26}$$

This variation of the couplings with energy scale may explain Dark Energy (Figure 2).

The original model assumed massless GUM Bosons, however Section 5.4 reviews an interesting application from Solid State Physics that reintroduces mass and leads to Equations (27) – (31). These mass scales may

**Table 22 – Corrected QSGUT Results**

| $N$ | $n^{\underline{th}}$ Force | $\alpha_n$ * | $n^{\underline{th}}$ Mass $\left(\mathrm{GeV}/c^2\right)$ | $n^{\underline{th}}$ Charge | $n^{\underline{th}}$ GUMB |
|---|---|---|---|---|---|
| 1 | Strong Nuclear | 1.63 to 68.1 | 0.134974 | Color | 8 gluons |
| 2 | Electro-magnetic | $7.297352568 \times 10^{-3}$ | 246 ***i*** | Electric | 1 photon |
| 3 | Weak Nuclear | $2.94812 \times 10^{-12}$ | 55.8 ***i*** | Weak Isospin | 3 IVB's |
| 4 | Gravity | $5.90603 \times 10^{-39}$ | $1.9 \times 10^{7}$ | Mass | 1-graviton |
| 5 | WIMP-Gravity | $1.34 \times 10^{-93}$ | $1.2 \times 10^{19}$ | Mass? | 3 WIMP-gravitons |

explain Inflation, Dark Matter and Higgs Theory. The corrected couplings and mass-scales are reviewed together in Table 22.

Note that a new force, WIMP-Gravity, and a new mass scale, $1.9 \times 10^4$ TeV / $c^2$ (the mass scale for WIMP-gravitons), are predicted. Although more forces may exist, the Planck mass scale may "pinch off" and suppress all higher $n$ vibrations of the string in four spacetime dimensions except the first five modes, thus allowing only five forces.

Chapter 7 represents the most speculative hypothesis implied by QSGUT, which combines basic truths from QSGUT and String Theory with a proposed 3-dimensional tetragonal Hyperflavor-Weak isospin **T** symmetry into a Group Theory approach to GUT with five forces and five generations (bosonic and fermionic vibrations of the string, respectively) contained in a 14-dimensional (spacetime plus two 3-branes plus two 2-branes) Supersymmetric SUSY $SU(15)$ Lie algebra GUT or a 12-dimensional SUSY $E12$ TOE that connects with String Theory on the high-energy end of the spectrum; then condenses down into the Standard Model, Gravity, and possibly sequestered remnants of Supersymmetry (SUSY), Hyperflavor, and WIMP-Gravity at low energies. Hyperflavor contains *SO*(8) subsets of Hyperflavor Leptons $\left(e, \nu_e, \mu, \nu_\mu, \tau, \nu_\tau\right)_{L,R;\ 1-7}$, Quarks $\left(u, d, c, s, t, b\right)_{c;\ L,R;\ 1-7}$ and Leptoquarks $\left(\mathrm{a}, \psi, \mathrm{v}, \omega\right)_{a;\ L,R;\ 1-7}$ (see Tables 7 and 11) that condense down to yield the observed patterns of Standard Model leptons and quarks in left-handed weak isospin doublets (see Table 8). We also have *SO*(8) subsets of bosons including 1) the HEW-Higgs boson sector; and 2) *Q, R, U* and *V* Grand Bosons that connect the five generations of fermions to each other, thus possibly explaining broken flavor symmetries, the CKM matrix and neutrino oscillations (see Tables 9 and 10) via the CKM-PMNS-Munroe Mixing Matrix.

Is there a Theory of Everything (TOE) that could describe all of these forces and interactions? This book has collected much evidence in favor of similarities amongst the forces, and given a clue that the TOE Equation may be a 14-dimensional SUSY *SU*(15) or a 12-dimensional SUSY *E*12 Lagrangian with proper field, matter and symmetry-breaking terms. In conclusion, we have covered topics ranging from $\alpha_1$ to $\omega_7$, introduced Hyperflavored GUM (bosons), and determined that the Universe is held together with strings and GUM (bosons)!

**Figure 4 – A Body-Centered Cubic Lattice of Fundamental Fermions in Hyperspace (See Front Cover)**

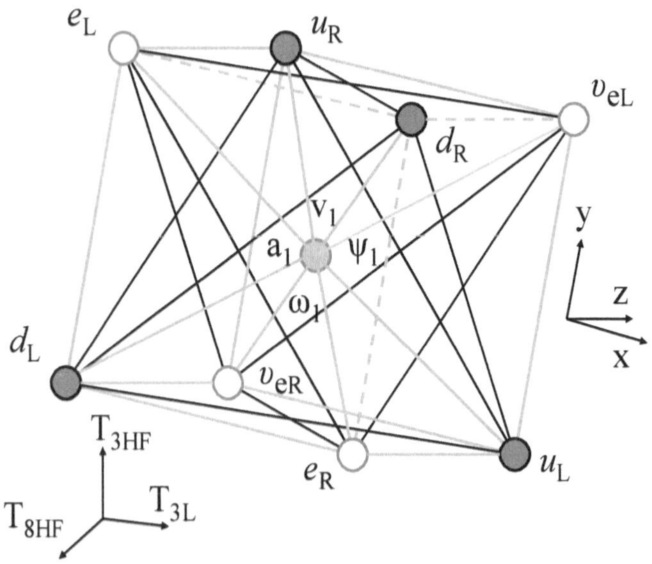

[1] Isaac Newton, in *Philosophiae Naturalis Principia Mathematica*, edited by A. Motte (University of California Press, 1966)

[2] James C. Maxwell, *Treatise on Electricity and Magnetism*, 3rd edition (Dover, 1954)

[3] A. Einstein, H. A. Lorentz, H. Minkowski, and H. Weyl, in *The Principle of Relativity, Collected papers*, edited by A. Sommerfeld (Dover, 1952); W. K. Clifford, Nature (London) **8**, 14 (1873).

[4] Sheldon Glashow, Nucl. Phys. **22**, 579 (1961); Steven Weinberg, Phys. Rev. Lett. **19**, 1264 (1967); A. Salam, in *Elementary Particle Theory*, edited by W. Svartholm (Almquist and Wiksell, 1968).

[5] U. Amaldi, W. de Boer, and H. Furstenau, Phys. Lett. B **260**, 447 (1991); P. Langacker and M. Luo, Phys. Rev. **D44**, 817 (1991); J. Ellis, S. Kelley, and D. V. Nanopoulos, Phys. Lett. B **260**, 131 (1991).

[6] E. Schrodinger, Ann. Phys. (Leipzig) **79**, 361 (1926).

[7] R. Balescu, *Equilibrium and Nonequilibrium Statistical Mechanics* (John Wiley and Sons, 1975), p.37.

[8] M. Kaku and J. Trainer, *Beyond Einstein - The Cosmic Quest for the Theory of the Universe* (Bantam Books, 1987), p.160.

[9] Lord Rayleigh, Philos. Mag. **49**, 539 (1900); M. Planck, Ann. Phys. (Leipzig) **4**, 553 (1901); J. H. Jeans, Philos. Mag. **10**, 91 (1905).

[10] "Review of Particle Physics", Particle Data Group, S. Eidelman *et al.*, Phys. Lett. **B592**, (2004), p.1.

[11] P. D. B. Collins, A. D. Martin, and E. J. Squires, *Particle Physics and Cosmology* (John Wiley and Sons, 1989), p.42.

[12] Paul A. M. Dirac, Nature (London) **139**, 323 (1937); Paul A. M. Dirac, Proc. R. Soc. London, Ser. A **165**, 199 (1938).

[13] E. C. J. Stuekelberg and A. Peterman, Helv. Phys. Acta **26**, 499 (1953); M. Gell-Mann and F. E. Low, Phys. Rev. **95**, 1300 (1954).

[14] W. K. Rose, *Astrophysics* (Holt, Rinehart and Winston, 1973), p.56.

[15] P. Atkins, *Physical Chemistry*, **6**th ed. (Oxford University, 1997), p.215.

[16] C. E. Shannon (ed.), A. D. Wyner (ed.), and N. J. A. Sloane, Claude E. Shannon: Collected Papers, Wiley-IEEE Press (1993).

[17] Ian Hinchliffe, http://www-theory.lbl.gov/~ianh/alpha/alpha.html, 2002.

[18] P. W. Higgs, Phys. Lett. **12**, 132 (1964); P. W. Higgs, Phys. Lett. **13**, 508 (1964); P. W. Higgs, Phys. Rev. **145**, 1156 (1966).

[19] J. Ellis and G. G. Ross, Phys. Lett. B **117**, 397 (1982); J. M. Frere and G. L. Kane, Nucl. Phys. B **223**, 331 (1983).

[20] N. W. Ashcroft and N. D. Mermin, *Solid State Physics* (Saunders College, 1976), p.225-229.

[21] A. H. Guth, Phys. Rev. **D23**, 347-356 (1981); A. Guth and P. Steinhardt, "The Inflationary Universe", *Scientific American*, May 1984, p. 116-128.

[22] S. Weinberg, *Gravitation and Cosmology: Principles and Applications of the General Theory of Relativity* (John Wiley and Sons, 1972), p.151-163.

[23] F. Halzen, R. A. Vasquez, T. Stanev, and H. P. Vankov, Astropart. Phys. **3**, 151-156 (1995).

[24] S. Hawking, *A Brief History of Time - From the Big Bang to Black Holes* (Bantam Books, 1988), p.96-97.

[25] KASCADE Collaboration, K.-H. Karnpert *et al.*,"Cosmic Ray Spectra and Mass Composition at the Knee", Nucl. Phys. Proc. Suppl. **136**, p.273-281 (2004).

[26] J. D. Barrow and J. K. Webb, "Inconsistent Constants", *Scientific American*, June 2005, p. 57-63.

[27] R. Chiao, W. Fitelson and A. Speliotopoulos, "Search for Quantum Transducers Between Electromagnetic and Gravitational Radiation: A Measurement of the Upper Limit on the Transducer Efficiency of Yttrium Barium Copper Oxide", Apr. 2003, arXiv:gr-qc/0304026v1.

[28] E. Podkletnov and G. Modanese, "Impulse Gravity Generator Based on Charged Y Ba(2) Cu(3) O(7-Y) Superconductor with Composite Crystal Structure", Aug. 2001, arXiv:physics/0108005v2.

[29] B. S. DeWitt, Phys. Rev. Lett. **16**, 1092 (1966).

[30] A. Reiss and M. Turner, "From Slowdown to Speedup", *Scientific American*, Feb. 2004, p. 62-67.

[31] A. I. Shlyakhter, Nature **264**, 340 (1976); Yasunori Fujii, Lect. Notes Phys. **648**, 167-185 (2004).

[32] P. D. B. Collins, A. D. Martin, and E. J. Squires, *Particle Physics and Cosmology* (John Wiley and Sons, 1989), p.158-186.

[33] Y. Fukuda *et al.*, Phys. Rev. Lett. **82**, 1810-1814 (1999).

[34] H. Georgi, Nucl. Phys. B **156**, 126-134 (1979).

[35] N. Cabibbo, Phys. Rev. Lett. **10**, 531 (1963); M. Kobayashi and T. Maskawa, Prog. Theor. Phys. **49**, 282 (1972).

[36] P. Atkins, *Physical Chemistry*, **6**th edition (Oxford University Press, 1997), p.432, 954.

[37] N. W. Ashcroft and N. D. Mermin, *Solid State Physics* (Saunders College, 1976), p.68-79.

[38] First evidence of the 2nd Generation (muon in cosmic rays): C. D. Anderson and S. H. Neddermeyer, Phys. Rev. **50**, 263 (1936). First evidence of the 3rd Generation (CP Violation explained by the CKM Matrix): J. H. Christenson *et al.*, Phys. Rev. Lett **13**, 138 (1964), M. Kobayashi and T. Maskawa, Prog. Theor. Phys. **49**, 282 (1972).

[39] ALEPH Collaboration, D. Decamp *et al.*, Phys. Lett. **B253**, 399 (1990).

[40] M. Kaku and J. Trainer, *Beyond Einstein - The Cosmic Quest for the Theory of the Universe* (Bantam Books, 1987), p.13.

[41] L. Randall, *Warped Passages – Unraveling the Mysteries of the Universe's Hidden Dimensions* (Harper Perennial, 2006), p. 335.

[42] V. Stenger, Eur. J. Phys. **11**, 236 (1990).

[43] V. D. Barger and R. J. N. Phillips, *Collider Physics* (Addison-Wesley, 1987), p.467.

[44] B. Pontecorvo, JETP (USSR) **37**, 1751-1757 (1959); Z. Maki, M. Nakagawa and S. Sakata, Prog. Theor. Phys. **23**, 1174-1180 (1962).

[45] N. W. Ashcroft and N. D. Mermin, *Solid State Physics* (Saunders College, 1976), p.91-92.

[46] E. Witten, Nucl. Phys. B **443**, 85-126 (1995).

[47] A. Garrett Lisi, "An Exceptionally Simple Theory of Everything", Nov. 2007, arXiv:0711.0770v1.

[48] For a review of the Anthropic Principle, see I. G. Barbour, Religion and Science – Historical and Contemporary Issues, p. 204-209, Harper Collins (1997).

www.ingramcontent.com/pod-product-compliance
Lightning Source LLC
Chambersburg PA
CBHW022104170526
45157CB00004B/1475
* 9 7 8 1 4 3 5 7 0 6 9 8 9 *